CAUGHT IN THE NET

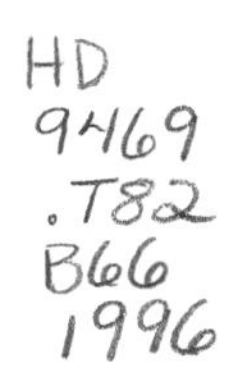

CAUGHT IN THE NET
THE GLOBAL TUNA INDUSTRY, ENVIRONMENTALISM, AND THE STATE

ALESSANDRO BONANNO AND
DOUGLAS CONSTANCE

UNIVERSITY PRESS OF KANSAS

Published by the University Press of Kansas (Lawrence, Kansas 66049), which was organized by the Kansas Board of Regents and is operated and funded by Emporia State University, Fort Hays State University, Kansas State University, Pittsburg State University, the University of Kansas, and Wichita State University

Library of Congress Cataloging-in-Publication Data

Bonanno, Alessandro.
Caught in the net : the global tuna industry, environmentalism, and the state / Alessandro Bonanno and Douglas Constance.
p. cm.
Includes bibliographical references and index.
ISBN 0-7006-0738-2 (cloth) ISBN 0-7006-0739-0 (pbk.)
1. Tuna industry—Government policy. 2. Dolphins—Mortality. 3. Tuna fishing—Environmental aspects. 4. Purse seining—Environmental aspects. I. Constance, Douglas. II. Title.
HD9469.T82B66 1995
338.3′72758—dc20
95-31617

British Library Cataloguing in Publication Data is available.

Printed in the United States of America

10 9 8 7 6 5 4 3 2 1

The paper used in this publication meets the minimum requirements of the American National Standard for Permanence of Paper for Printed Library Materials Z39.48-1984.

CONTENTS

ACKNOWLEDGMENTS

During the three-year period necessary for the completion of this volume, we had the opportunity to count on the assistance and support of a significant number of friends and colleagues. We would like to thank those who participated in the various conferences, meetings, and workshops organized to study the globalization of the agricultural and food sector. It was in this context that the idea of writing a book on the tuna-dolphin controversy developed, and it was there, above all, that we had the opportunity to discuss various aspects and phases of our work. Among those who commented on our endeavor we would like to thank Bill Friedland, Bill Heffernan, Terry Marsden, Lourdes Gouveia, Luis Llambí, Alberto Arce, Bob Schaeffer, Marie Christine Renard, Mary Hendrickson, Monica Bendini, Marta Palomares, Joe Molnar, David O'Brien, Craig Harris, Frank Vanclay, Nick Paretsky, Nelson Prato, Enzo Mingione, Enrico Pugliese, Karen Bradley, Mara Miele, and Victoria Godino. Particular thanks go to Professor Robert J. Antonio for his insightful comments on the issue of tuna loins. Special thanks also go to the participants in the seminars "Globalization of the Agricultural and Food Sector," held in Caracas, Venezuela, in May 1992, and "Sociology of Agriculture," held at the University of Comahue, Argentina, in November 1993. Additionally, our gratitude goes to the members of the Department of Social Sciences of the University of Messina, Italy, for their comments during the final stages of this book.

Despite the contributions of the friends mentioned above, the responsibility for the content of this volume rests exclusively with us.

Moreover, the listing on the authors' names is strictly alphabetical. This book is the outcome of a full cooperation between equals, and neither of the authors is or should be considered the major contributor to or first author of it.

ABBREVIATIONS

BANPESCA	Banco Nacional Pesquero y Porturio S.N.C.
CHB	California Home Brands
CITES	Conference on International Trade in Endangered Species of Wild Flora and Fauna
CYRA	commission's yellowfin regulatory area
DWFN	distant water fishing nations
EC	European community
EEC	European Economic Community
EEZ	exclusive economic zone
EII	Earth Island Institute
EPA	Environmental Protection Agency
ETP	Eastern Tropical Pacific
FAO	Food and Agriculture Organization
FDIC	Federal Deposit Insurance Corporation
FFA	Forum Fisheries Agency
FSLIC	Federal Savings and Loan Insurance Corporation
GAO	General Accounting Office
GATT	General Agreement on Tariffs and Trade
GM	General Motors Corporation
GOP	globalization of production
GRREC	Groupe de recherche sur la regulation d'economies capitalistes
IATTC	Inter-American Tropical Tuna Commission
IDCA	International Dolphin Conservation Act

IMF	International Monetary Fund
LDCs	less-developed countries
MARAD	Maritime Administration, U.S.
MERCOSUR	Economic Market of the South
MMPA	Marine Mammal Protection Act
MNCs	multinational corporations
MSP	maximal sustainable population
NAFTA	North American Free Trade Agreement
NATO	North Atlantic Treaty Organization
NICs	newly industrialized countries
NIDL	new international division of labor
NMFS	National Marine Fisheries Service
NSM	new social movements
OECD	Organization for Economic Co-operation and Development
OPEC	Organization of Petroleum Exporting Countries
OSP	optimum sustainable population
PPM	Productos Pesqueros Mexicanos
RAN	resource adjacent nations
SALs	structural adjustment loans
SEPESCA	Subsecretario de Pesca
SPF	South Pacific Forum
SSA	social structures of accumulation
TFC	transnational financial capital
TNCs	transnational corporations
USDA	U.S. Department of Agriculture
USITC	U.S. International Trade Commission
WIPO	World Industrial Patent Organization
WTP	Western Tropical Pacific

1
INTRODUCTION

Halfway out of the water, a Costa Rican spinner dolphin, caught by its beak in the mesh, wriggles free and drops back in. The triangular fin cuts under, and the dolphin rejoins its mates in the steadily shallowing belly of the net. Another entangled spinner rises. Its head is pressed awkwardly forward, its dorsal fin bent sideways, its beak half open. It is nearly to the power block when something—the dorsal?—appears to break, and the dolphin and dark fragments of it tumble back into the sea. The net at first is selective. The youngest spinners, quickest to tire, are the first to be caught in it. For a time mostly the slender bodies of calves are borne up the conveyer of the red mesh. The camera rolls on and soon the adults, too, die, or surrender and begin the climb. They rise, a dense mass of bodies, until the steepening angle of the net tips them off, four or five at a time, to pitch downward, beak over tailfin, to be caught again by the net's shallower angle at the waterline, to begin the climb once more.

A large male dolphin, completely enshrouded in mesh, approaches the power block. It twists and bucks wildly. The camera notices. The struggles of most dolphins at this stage are much feebler. In this animal the life force seems unusually strong. The camera zooms in. The dolphin passes quickly through the power block. Emerging, it slides down the red mound of braided net, shoved and guided by the hands of fishermen. A strange thing has happened to it. The amplitude of the big dolphin's struggles has flattened out. Where before its flukes traveled through a wide arc, a reflexive swimming motion,

> now they beat in a shallow, spasmodic flutter. That moment in the power block was too brief, it seems, to have wrought the change. But that is how it always seems, of course, to mortal creatures passing through that particular door. The fishermen slide the big dolphin, its flukes still fluttering, along the wet deck. They shove it to the top of a sluiceway and send it along to the sharks. (Brower describing a film by Sam LaBudde in Brower 1989, 45–46)

Sam LaBudde was a biologist with a strong wanderlust. After several fishery related jobs in Alaska, he went to San Francisco in the summer of 1987 to get a job with one of the "environmental heavyweights," thinking he wanted to work to save the rain forests. The Nature Conservancy was too "white-collar" and Greenpeace wanted him to canvass. At Earth Island Institute, a San Francisco umbrella organization, he picked up a publication of *Earth Island Journal* and read about the systematic killing of dolphins associated with tuna fishing. He asked the people at Earth Island Institute, David Phillips and Todd Steiner, why they weren't telling people about the atrocity. Phillips and Steiner replied that they were but they needed film of the killings, not just statistics. LaBudde, a former fisherman and a National Marine Fisheries Service (NMFS) observer, wondered if he could get on a tuna boat and thought he probably could (Brower 1989).

LaBudde talked the idea over with Phillips, Steiner, and Stan Minasian, the director of the Marine Mammal Fund. He also conferred with William Perrin, the NMFS biologist who had first brought the issue to public attention, and Perrin warned him that if he pursued his idea he might end up with "concrete galoshes." Although the environmental organizations did not have the money to finance such a venture, they told LaBudde that if he got aboard a tuna vessel, they would send him a video camera (Brower 1989).

In September 1987, LaBudde drove his Volkswagen to Ensenada, Mexico. It took him three weeks to get a job on the Panamanian tuna boat *Maria Luisa,* for his Spanish was bad and the tuna boat captains did not understand why an American fisherman wanted to work on a Mexican fishing boat. LaBudde explained that he was tired of the fast pace of U.S. life and wanted to learn Spanish. After finally getting a job, he called San Francisco and had a camera shipped to San Diego. The flight bringing in the camera was late, and after several taxi rides LaBudde arrived at the dock in Ensenada just eighteen minutes before the *Maria Luisa* was set to sail (Brower 1989). The important fruit of Sam LaBudde's wandering was the video of the dying dolphins. The video became a catalyst, and after years of pressure by environmental and consumer groups, the biggest tuna processors—Heinz (Star Kist), Van Camp (Chicken of the Sea), and Bumble Bee—announced that they would no longer accept tuna caught by "setting on dolphins."

What, then, is still so important about the tuna-dolphin controversy? Why is it so essential to devote a sociological book to this case? One answer is that the case provides a wealth of information concerning the characteristics and consequences of the current restructuring of the economy and society. The boycott of tuna caught by killing dolphins was just one moment in a long period of restructuring for the tuna industry, both in the United States and abroad, which has been paralleled by similar processes in a number of other economic sectors and subsectors. More specifically, the events of the case can be successfully employed to illustrate the shift from post–World War II capitalism, or Fordism, to a new and more-flexible form of capitalism whose emergence has been associated with the globalization of the economy and has frequently been called "global post-Fordism." "Perhaps even more important, the tuna-dolphin problem raises a complex series of sociological, and inevitably political, issues that include attitudes of and about fishermen, food processors and canners, retailers, animal-rights advocates, disparate groups of environmental activists, the popular press, and other news media" (National Research Council 1992, vii).

Fordism refers to the period between the end of World War I and the early 1970s. Furthermore, Fordism can be subdivided into low and high Fordism, the first indicating the period up to the end of World War II during which production processes were centralized and vertically integrated. There was the emergence and consolidation of large companies whose production facilities were concentrated geographically (i.e., in industrial cities such as Detroit, Manchester, and Milan) and economically (i.e., the rise and establishment of oligopolistic companies). High Fordism refers to the post–World War II years up to the beginning of the 1970s which were characterized by the expansion of oligopolies and their geographical concentration in advanced Western societies. The rapid economic growth experienced in those years also allowed mechanisms that redistributed wealth in ways never experienced before. Middle and lower classes benefited from the economic growth through stable employment, rising incomes, and the development of a welfare system. In return, conflict between labor and management was lessened by a corporatist pact of mutual nonaggression—the capital-labor accord.

Fordist rigidity pertains to the characteristic of the capital-labor accord. The technology of mass production required large-scale fixed investments and large structures, so factories were designed to manufacture uniform products over long production runs; they were not equipped to deal with sudden interruptions in the cycle of production such as those generated by lack of raw materials and/or, more important, by workers' strikes. Uniform production plus vulnerability to changes in the supply of raw materials and to labor actions define the meaning of Fordist rigidity.

Post-Fordism refers to the period comprising the middle and late portions of the 1970s and the following decades and involves a transition away from the Fordist system of production and the capital-labor accord. In essence, post-Fordism production units were decentralized into many localities and smaller factories while control still remained in the hands of oligopolies. Decentralization involved all facets of production including design, administration, and labor operations. Through decentralization the companies' ability to react to changes in the conditions of production was greatly enhanced, so labor actions and changes in the supply of raw material became less threatening. Also, companies could react more easily to the changes in consumer preferences and modified market conditions. Post-Fordism assumed a global dimension as decentralization came to involve the scattering of production locations around the globe. Products that once were manufactured under one roof (i.e., in one factory and in one locality) are now produced under many smaller roofs in many locations. As a result, the production of the different parts of a final product is delegated to dispersed units and locations.

Whereas Fordism is primarily characterized by national economies based on domestic mass production and consumption regulated by welfare-states, global post-Fordism is characterized by the internationalization of economic relations based on strategies of flexible accumulation and global sourcing and on the transformation of the nation-state. Flexible accumulation refers to the organization of production based on a multiplicity of labor arrangements, production strategies, and market access. Decentralized companies have a variety of possible arrangements from which they select the most appropriate options given overall economic conditions. For example, the automotive companies delegate the production of parts to smaller businesses in varied locations around the world and use subcontracting to ensure the timely delivery of standardized products, which are then assembled into the final product. Moreover, computerized technology means there can be rapid modifications of production to accommodate changes in consumer demands.

The selection of production sites and labor pools is not random. Rather, it is influenced by the companies' desire to utilize the most convenient factors of production, which are searched out and obtained globally. Global sourcing is the term for this specific process. For example, labor costs can be reduced by identifying and utilizing less-expensive labor pools, so production processes are decentralized to regions where labor costs are significantly lower. Similarly, stable social and labor relations may be established by identifying conditions around the world that are amenable to these objectives—such as a docile labor force and probusiness government policies. As indicated by a number of recent works, the transition from Fordism is so pervasive that it affects all sectors and aspects of life. The case of the tuna-dolphin controversy is only one episode that illustrates the characteristics and implications of this change.

Another reason for studying the tuna-dolphin controversy lies in the importance of the agro-food sector. By agro-food sector we refer to the complex of input industries, farming, fishing, processing, and distribution of food products. The strategic centrality of food has been well documented in the past. However, the current interest in this sector rests more in the fact that it is one of the most globalized, despite the common understanding that agro-food production is essentially local and exogenous to global circuits of production. In recent years, a wealth of studies have demonstrated this fact and, at the same time, have raised important questions concerning the ways in which global restructuring is carried out. For example, Tyson Food Inc., the largest poultry processor in the world, has implemented a production system whereby it sources low-cost Midwest grain to feed chickens in Arkansas. The breast meat is then removed for use by the U.S. fast food market. The remainder of the carcass is shipped to Mexico to be deboned by low-paid, female Mexican laborers. Finally, the "yakatori stick," marinated chicken on a bamboo stick, is exported to Japan where there is a high demand for marinated dark-chicken meat. Similarly, Cargill Inc., the largest agro-business corporation in the world, has slash-and-burn rain forest beef operations in Brazil, Mexico, and Honduras to supply the global demand for hamburgers. But it also has high-quality operations in the United States, Canada, and Australia (Heffernan and Constance 1994). This strategy of diversification allows Cargill to meet any type of demand for beef without depending on a particular production site and possible "social tension" that might hamper continuous supply. In line with these examples, this book continues a long tradition of concern about the structure of food production and economic concentration as they impact society.

Within the literature on sociology of the economy and society and the sociology of agriculture, there is an ongoing discussion regarding the characteristics of restructuring the socioeconomic system and the global food system (e.g., Bonanno et al. 1993, 1994; Friedmann and McMichael 1989; Gouveia 1994; Heffernan and Constance 1994; Kenney et al. 1989; Krebs 1992; Lobao 1990; McMichael and Myhre 1991; Sanderson 1986). This discussion finds increasing consensus in the fact that after the end of World War II there began a period of stable economic growth characterized by high levels of capital accumulation, legitimated by increased state intervention and abundance—the Fordist era (or mass production era or Social Democrat era). In the 1970s, a global accumulation crisis emerged, and the altered conditions demanded significant restructuring, which characterizes the current phase of capitalism, or global post-Fordism. Global post-Fordism, then, refers to attempts by corporations to implement strategies to restore accumulation and to diminish resistance to those strategies (Bonanno 1991; Friedland 1991b; Friedmann and McMichael 1989; Gordon et al. 1982; Lash and Urry 1987; McIntyre 1991; Sassen 1990). The method includes the use of much more flexible forms of production than those utilized in Fordism and the capacity to search worldwide (i.e., global sourcing) for more convenient production factors in order to avoid high labor costs, labor resistance, proenvironmental legislation, and other forms of state regulation characteristic of welfare-states under the Fordist era (Constance and Heffernan 1991; Friedmann and McMichael 1989; Harvey 1990; Sanderson 1986).

Business' strategy manifests itself as deindustrialization in First World states and capital flight to less-regulated newly industrialized states (NICs) and/or Third World states (Bluestone and Harrison 1982; Harrison and Bluestone 1988; Lash and Urry 1987; Lipietz 1987a, 1987b). In general, deindustrialization is characterized by two complementary phenomena, the decentralization of production and the informalization of labor. The decentralization of production implies movement of production facilities away from large, centralized, union-based production sites in the First World to fragmented production arrangements based on subcontracting and joint ventures in NICs such as Brazil, Korea, Mexico, and Taiwan and/or Third World sites. The informalization of labor implies more women and minority persons employed in a labor process characterized by flexible and temporary jobs with low wages and no benefits (Bowles et al. 1988; Bluestone and Harrison 1982; Gordon et al. 1982; Harrison and Bluestone 1983, 1988; Lash and Urry 1987; Lobao 1990: Mingione 1991; Newman 1988; Piore and Sabel 1984; Sassen 1990). These phenomena, which are referred to as the globalization of production, have resulted in a new international division of labor that is based on global sourcing and characterized by manufacturing and processing in the NICs or Third World (or in the First World and NICs by Third World immigrants), consumption in the First World, and growing middle classes in the NICs (Harvey 1990; Lipietz 1987b; Sanderson 1985, 1986).

In response to capital flight, First World states deregulated or reregulated their economies in an attempt to maintain and attract investment and thereby avoid disinvestment and deindustrialization (Bluestone and Harrison 1982; Bowles et al. 1988; Friedland 1994; Lash and Urry 1987; Lipietz 1987b). Such actions imply a weakening of the nation-states' ability to effectively coordinate domestic accumulation of capital and its legitimation to the broader citizenry as well as to protect labor from informalization. These limits on the nation-state's abilities are primarily due to the globalization of economic processes and the growth of powerful transnational corporations (TNCs) (Aglietta 1979; Bonanno and Constance 1993; Friedmann and McMichael 1989; Harvey 1990; Lash and Urry 1987; Lipietz 1987b; McMichael 1991; McMichael and Myhre 1991; Sanderson 1986). More specifically, the nation-state's limited ability to coordinate national accumulation and legitimation is due to the separation or dislocation between the sphere of influence of the state—i.e., the nation-centered sphere—and the sphere of influence of transnational capital—i.e., the globalized production sphere—resulting in the crisis of the nation-state in the post-Fordist transition (Bonanno 1994; Lash and Urry 1987; McMichael and Myhre 1991). The assumed transformed character of the nation-state, its supposed weakening, its relationship with

TNCs and other social factors such as the labor and environmental movements, and the overall relevance of the state in global post-Fordism are all foci of this book.

The success of the Fordist regime of accumulation was based on stable, neo-corporatist national arrangements between labor and capital that were mediated and enforced by the welfare-state—under Fordism national capitals were aligned with nation-states. This system was predicated on the extraction of low-cost natural resources and labor from the Third World by First World multinational corporations (MNCs). The growth of the global economy and the emergence of powerful TNCs have destabilized or destroyed these corporatist arrangements, and the old class compromises, such as the capital-labor accord, were broken in an attempt to resolve the global accumulation crisis of the early 1970s.

Under Fordism, Keynesian economic policies designed to manage recessionary tendencies were combined with progressive social-democratic institutions such as legitimized unions, unemployment insurance, and minimum wage laws to facilitate a balance between mass production and mass consumption. The managed balance, together with the economic efficiencies of Taylorist mass production and a credit system that enabled the workers to purchase the fruits of their labor (i.e., cars, appliances, and houses), provided intensive levels of accumulation that lasted into the 1960s (Aglietta 1979; Lash and Urry 1987; Lipietz 1987b). The U.S. state, the state apparatus of the United States, was a critical component of Fordism in the United States, with the New Deal serving as the prime example. According to Aglietta, in the Fordist era, "the state becomes the bearer of the monetary constraint" (1979, 349). Regulating a national money through a central bank, mediating a class compromise between capital and labor, providing social security and unemployment insurance, and providing subsidies and protection to farmers all indicate the vital role that the U.S. state played as part of the Fordist regime.

In the 1960s Fordism started to unravel as the cost of production began to rise in the industrialized world. The technical efficiencies associated with Taylorist/Fordist mass production were exhausted. Fixed costs per capita were increasing, and the cost of the social wage was increasing faster than the increase in profit. Labor had been successful in using the welfare-state to extract numerous wage and benefit concessions from management, and labor was also beginning to rebel against the dehumanizing aspects of Taylorist mass production. Other state regulations such as antitrust and environmental policies increased the cost of production in the First World welfare-states (Aglietta 1979; Gordon et al. 1982; Lash and Urry 1987; Lipietz 1987b; Piore and Sabel 1984).

The cost of production also increased due to resistance emerging from developing nations. National liberation movements, anti-Western movements such as Islamic fundamentalism, and even moderate expressions of dissent with traditional Western international policies such as (OPEC) generated profound alterations in the patterns of transfer of wealth from developing to developed regions. Developing countries began to challenge the dominance or hegemony of the United States in world markets. During the 1960s NICs began to penetrate the markets dominated by the industrialized nations during the Fordist era by combining Taylorist mass production with lower labor costs and few governmental constraints on business. Lipietz (1987a) refers to this as peripheral Fordism. The stable international monetary regime facilitated by Bretton Woods, the International Monetary Fund (IMF), and the General Agreement on Tariffs and Trade (GATT)—which produced and supported the dominance of the U.S. dollar in international markets—collapsed as the dollar was moved off the gold standard and onto a variable rate system. The oil shocks and massive Soviet grain purchases of the 1970s destabilized national economies, and the hegemonic position of the United States was successfully challenged by Japan and the European Economic Community (EEC) (Aglietta 1979; Bluestone and Harrison 1982; Gordon et al. 1982; Lash and Urry 1987; Lipietz 1987b; Piore and Sabel 1984).

Although the Fordist era had been based on national mass production balanced with national mass consumption and regulated by a relatively autonomous welfare-state, the transition to post-Fordism was marked by vigorous attempts by large MNCs to penetrate each other's First World markets. Large MNCs also reacted to the increased cost of production by restructuring, which included decentralization through the transfer of some production to NICs or the Third World, where labor was more docile and state regulations were minimal. This early transitional phase was characterized by the multinational phase of capitalism where large First World firms were still "attached" to their nation-states (Aglietta 1979; Lash and Urry 1987; Lipietz 1987b; McMichael 1991; McMichael and Myhre 1991; Piore and Sabel 1984).

The deregulation of the global finance system associated with the demise of Bretton Woods globalized capital circuits. As MNCs cut their attachments to their traditional nation-states, they became transnational corporations with operations in many countries and diffused allegiances. For example, Cargill's operation covers forty-nine countries with more than 800 offices and plants employing more than 55,000 workers and trades in more than 100 different commodities (Constance and Heffernan 1991; Kneen 1990). As TNCs began to source financing, production, and markets globally, the nation-state faced increasing difficulty in regulating TNCs domestically. Kneen's study (1990) of the effects on Canadian agriculture of Cargill's emergence as a grain TNC is one of the best examples of this phenomenon. Regulations that increased production costs could provoke capital flight that would imply lost jobs and lowered standards of living for workers in those industries (Harrison and Bluestone 1988; Lash and Urry 1987; Piore and Sabel 1984). Thus, the post-Fordist era was the transition from the national and international relations associated with the Fordist era of postwar U.S. hegemony to the global economy characterized by interrupted and increasingly uneven accumulation (MacEwan and Tabb 1989; Gottdiener and Komninos 1989), by state fiscal crises (O'Connor 1973, 1974, 1986) and by a realignment of economic power (Friedland 1991a). The corporatist arrangements within the Western welfare-states begin to dissolve as TNCs avoid investing in highly regulated states.

Chapter 2 of this book consists of an extensive review of literature on the transition from Fordism to global post-Fordism. An overview of the characteristics of Fordism and global post-Fordism is presented, along with a critical examination of a number of works that directly or indirectly address the issue of the move away from Fordism. Chapter 3 continues the review of the literature, focusing on the issue of the state, its centrality in Fordism, and its continued importance in post-Fordism. Various theories of the state are reviewed, including those arguing the emergence of a transnational state, along with considerations on the reconceptualization of the state in a globalized system. Chapters 4, 5 and 6 present the case study. The seventh and final chapter links the evidence presented in the illustration of the case with salient issues introduced in the literature review.

In Chapter 7 we argue that global post-Fordism is characterized by increased flexibility in crucial spheres such as employment, use of labor, and the production process. Additionally, we indicate that flexibility should not be equated with disorganized capitalism but rather signifies the creation of a new organization based on a more sophisticated system of options available to transnational players. We maintain that both TNCs and other groups in society have some interest in desiring a transnational form of state, but the type of state preferred by these groups differs significantly. The chapter concludes by an exploration of the relationship between TNCs and other social actors. Of significant importance is the fact that, in the context of the case, environmentalists and workers oppose each other. This situation creates a split among forces that are potentially at odds with TNCs but that ultimately do not challenge the TNCs' ability to operate freely in the global system. Simultaneously, it would be erroneous to conclude that TNCs do not encounter opposition to their restructuring strategies. Indeed, the environmentalists' twenty-year struggle constituted a formidable opposition to corporate actions designed to avoid pro-dolphin legislation; in the end the environmentalists won. But the fragmentation of opposition to dominant players (i.e., the inability of environmentalists and workers' organizations to work together against TNCs), the chapter concludes, poses the crucial potential for the creation of new forms of solidarity among subordinate groups.

Before concluding this introduction, we believe that it is important to provide a simplified overview of the case of the tuna-dolphin controversy in preparation for the detailed exposition carried out later in the book. Substantively, the case reveals that TNCs use the strategies of flexible accumulation to avoid state regulations, which in turn limits the state's ability to effectively regulate and assist TNCs' actions. Moreover, as a consequence of flexible accumulation, the state's ability to organize its national economy and guarantee social rights in the form of workers' rights and environmental protection to its citizens is also diminished.

The tuna-dolphin controversy began in the late 1950s when U.S. tuna fishermen based in San Diego responded to increased rates of tuna boat seizures resulting from territorial fishing water disputes and to Japanese incursions into the U.S. tuna market by developing a new tuna fishing technology based on large purse-seine nets. Purse-seine tuna fishing substituted capital for labor by replacing the labor-intensive practice of several fishermen using baited poles with large tuna boats utilizing nets up to a mile long. The new technology facilitated huge catches of tuna in a single setting of the net—and killed hundreds of thousands of dolphins per year in the process.

During the 1960s environmentalists and conservationists declared this fishing technique, along with other hunting and fishing techniques such as clubbing baby harp seals, illegitimate. They pressured the U.S. state into passing legislation in 1972—the Marine Mammal Protection Act (MMPA)—that protected and managed the harvest of marine mammals. Specifically, MMPA ordered incidental dolphin kills associated with tuna fishing to be reduced to "insignificant levels approaching 0." MMPA was contested from its inception by the tuna industry and supporting portions of the U.S. state, most notably the Department of Commerce and its subdivision, the National Marine Fisheries Service, which was mandated to implement MMPA. The tuna industry was initially granted a two-year grace period to develop new dolphin-safe fishing techniques, but no new techniques were forthcoming and dolphin kills remained high.

Environmentalists then used sympathetic courts to enforce a decreasing quota system to reduce dolphin kills systematically. Although the quota system initially did bring down the number of dolphin kills, under the Reagan administration the original intent of MMPA to reduce dolphin kills to insignificant levels approaching zero was temporarily abandoned. Because of lobbying by the tuna industry and sympathetic support from the Reagan administration and the Department of Commerce, "0" was redefined in 1984 as "20,500."

Amendments to MMPA in 1984 required the Department of Commerce to ban imports of tuna from foreign fishing fleets that did not have dolphin kill rates comparable to the U.S. standard. By 1988 the National Marine Fisheries Service (NMFS) under the Department of Commerce had not produced the necessary data to measure comparability. The delaying action prompted the environmentalists to again enlist the support of the courts to force the Department of Commerce to obey U.S. law. During the 1988 MMPA reauthorization hearings, NMFS and the Department of Commerce came under strong criticism from Congress for their foot-dragging regarding implementation of MMPA.

Disillusioned by the lack of progress in the legislative arena, in 1988 environmentalists launched a consumer boycott of the three major tuna processors in the United States (Heinz's Star Kist, Ralston Purina's Chicken of the Sea, and Pillsbury's Bumble Bee), which controlled about 70 percent of the U.S. market. In 1989 the environmentalists garnered further support in Congress and proposed the Dolphin Protection Consumer Information Act to counter the continued use of purse-seine nets.

In 1990 the three largest tuna processors agreed voluntarily to accept only dolphin-safe tuna.

In an attempt to bypass the regulations of the U.S. state and combat foreign competition, the U.S. tuna industry restructured with a two-point strategy. First, a large percentage of American tuna boats reflagged under foreign nations to avoid U.S. regulations requiring observers on tuna boats to document dolphin kills. Second, two of the top three tuna processors (Ralston Purina and Pillsbury) sold out to Asian companies to avoid the negative publicity of the consumer boycott and the higher cost of business due to MMPA and increased foreign competition. In effect, the industry moved around the regulations.

Although in 1988 the Inter-American Tropical Tuna Commission (IATTC) implemented an observer program for the foreign tuna fleets to provide comparability data, by early 1990 NMFS still had not provided any data for comparability figures, and the majority of dolphin kills were now attributable to foreign fleets. Again the environmentalists used the courts to impose an immediate embargo on the foreign fleets until proof was provided regarding reduced dolphin kills. NMFS and the Department of Commerce lost their appeal of the embargo, which mostly affected Mexico and Venezuela, whose fleets had expanded their tuna catch as the U.S. fleets decreased theirs.

The embargo was reluctantly enforced by the Bush administration, which was negotiating the North American Free Trade Agreement (NAFTA) with Mexico. U.S. sanctions on Mexico regarding environmental issues were an embarrassment to the Bush administration as it tried to get fast-track authority for NAFTA. Both the Salinas and Bush administrations argued that the tuna-dolphin controversy should be handled outside the NAFTA talks in international forums. The Bush administration tried to resolve the situation with bilateral negotiations but failed, while Mexico filed suit under GATT and won a favorable decision that food imports could not be banned because of their production method.

The GATT ruling raised questions about several international treaties and accords to protect the environment and prompted more bilateral attempts to resolve the controversy. In 1991 Mexico agreed to defer action on the GATT complaint in exchange for a promise by the Bush administration to work to convince Congress to change MMPA regulations. Mexico also issued a plan to reduce dolphin kills and hired U.S. public relations firms to advertise the plan in prominent American newspapers. The proposed changes in MMPA faced overwhelming opposition in Congress and were unacceptable to the environmentalists.

In early 1992 the environmentalists again used the courts to secure compliance with MMPA. As a result of the first embargo on Mexico and Venezuela, tuna was increasingly being transshipped to third-party countries and then imported into the United States. A secondary embargo was enacted to ban the import of "laundered" tuna from about twenty countries that bought tuna from Mexico or Venezuela. EEC officials and other countries launched another complaint under GATT, accusing the United States of unfair trade practices.

Later in 1992 the United States, Mexico, and Venezuela reached a preliminary agreement to protect dolphins from purse-seine nets via a five-year moratorium. At the same time, IATTC negotiated an accord with the major tuna fishing nations to again set up a declining quota system to systematically reduce dolphin kills to "insignificant levels approaching 0." Environmentalists argued that neither program was acceptable if it included the lifting of the trade sanctions.

Finally, in late 1992 an agreement between the United States, Mexico, and Venezuela was reached. The compromise contained the five-year moratorium included in previous compromises but also included an agreement by Mexico and Venezuela to face import embargoes on all seafood except shrimp if they failed to meet the conditions of the accord. The inclusion of this last provision received the support of the environmentalist lobby. The International Dolphin Conservation Act of 1992 (IDCA) was signed by President Bush in October 1992. Although the issue seems resolved, the Mexican government has yet to sign the agreement and the EEC and other countries still have a complaint before GATT.

In essence, the ownership and geopolitical structure of the tuna industry was transformed to circumvent the costly regulations mandated by MMPA. Neither the United States or Mexico or Venezuela could secure tuna processing jobs, as TNCs relocated their processing operations to avoid MMPA and counter rising international competition. Reconciliation of the tuna-dolphin controversy was, and still is, extremely problematic for the U.S. state and other states. Environmentalists effectively used the judicial system to force implementation of MMPA, while the tuna industry used the Commerce Department to stall that implementation. The challenge under GATT from the EEC and other nations still calls into question the longevity of MMPA.

Finally, the case illustrates the end of the Fordist period of tuna fishing and the emergence of a complex post-Fordist period. The former was based on a vertically integrated and nationally oriented, rigid form of production, i.e., national mass production matched to national mass consumption regulated by an interventionist state. Fordist tuna fishing allowed the rapid accumulation of capital, the expansion of stable markets, and the creation of dependable labor pools. This system was challenged by U.S. environmental groups that were backed by fractions of the domestic state, such as the judiciary system and segments of the legislative branch of government, and by actions of developing countries that were attempting to create their own Fordist tuna production systems. TNCs responded to rising costs by decentralizing production away from the United States and by taking advantage of lower labor costs and more favorable business regulations abroad. This strategy included, among other things, reflagging of tuna boats, shifting canning operations offshore, and introducing new processing technology. These events brought in new players such as foreign fishermen, developing nation-states, and international institutions such as the IATTC, NAFTA, and GATT. Decentralization and informalization of operations, intervention of transnational players, and the crisis of domestic forms of production control all represent fundamental elements of the transition to post-Fordism. Although the tuna-dolphin controversy is an ongoing affair, our analysis terminates at the end of 1992. There have been a few developments since then; however, they have not altered the substance and direction of the case.

2
THE DEBATE ON THE TRANSITION FROM FORDISM TO GLOBAL POST-FORDISM

At the outset of this chapter we present our interpretation of the transition from Fordism to global post-Fordism. This analysis is based on our reading of current literature and knowledge of data on relevant socioeconomic trends and is followed by a review of salient literature on the topic. Both our reading of the transition to global post-Fordism and the review of the theses of other authors constitute the overall framework for the interpretation of the case study and for the reading of the concluding chapter. The literature presented below is grounded in a broader Marxian framework and includes works by O'Connor (1989); European regulationists such as Aglietta (1979, 1982) and Lipietz (1987a, 1989); British authors Lash and Urry (1987), Harvey, (1990) and Jessop (1990); and American radicals such as Gordon (1978, 1980, 1988), Gordon et al. (1982), Piore and Sabel (1984), and Sabel (1982). The foci of these works are the characteristics of mature capitalism, its crisis in the late twentieth century, and the transition from Fordism to global post-Fordism. This debate is largely motivated by the exhaustion of emancipatory expectations contained in the Fordist era and the emergence of a process of rethinking that involves developmental strategies, social institutions and groups, and, above all, the cultural milieu in which hopes for better future arrangements were lodged.

Our opening analysis draws on many of the arguments presented in these works, but differs from them in significant ways. Our view of Fordism and global post-Fordism is not grounded in regulationist epistemological assumptions, though it reaches a number of similar conclusions. Additionally, it rejects the argument of disorganized capitalism, made by scholars within the British school. Instead we suggest the emergence of new forms of organization of the production process. Also rejected is the thesis of the "long waves" favored by some American radicals. Our reading of recent socioeconomic history insists on the emergence in the early 1970s of a distinct phase of capitalism that is qualitatively different from that characterizing the accelerated socioeconomic growth following World War II. Finally, although we do not reject arguments about "spatio-temporal compression" and "turnover time," we interpret the generating conditions of post-Fordism in a broader, multicausal manner.

After we present our analysis of the transition from Fordism to global post-Fordism, we review the literature indicated above, beginning with James O'Connor's summary analysis of various interpretations of the current restructuring of mature capitalism and continuing with the European debate that is the source of both concepts. Attention is paid to the works of the Parisian regulationist school, headed by Michel Aglietta and Alain Lipietz, which argues the centrality of national social formations in the emergence of global capitalism. Also discussed is the British perspective, exemplified by the works of Scott Lash, John Urry, David Harvey, and Bob Jessop, which is centered on the debate on the proposed disorganization of capitalism under post-Fordism. The American debate among radical writers focuses on the rejection of the epochal fracture between Fordism and post-Fordism and the emancipatory aspects embedded in post-Fordist flexibility. In this section we rely on the works of David Gordon, Richard Edwards, Michael Reich, Michael Piore, and Charles Sabel.

FROM FORDISM TO GLOBAL POST-FORDISM

Changes in the organization of the economy and society have prompted sociologists and other social scientists to argue the transition from Fordism to a post-Fordist system (Friedmann and McMichael 1989; Kenney et al. 1989; Lash and Urry 1987; Lipietz 1987a, 1987b; Lobao 1990; Piore and Sabel 1984). As illustrated by Antonio and Bonanno (1995), Fordism refers to the type of capitalism that began to emerge in the later nineteenth century, was consolidated after World War I, and reached its peak in the post–World War II era. The concept was inspired by Henry Ford's assembly-line operation, which seemed to realize Frederick Taylor's hopes about applying science in production and administration to improve collectively the conditions of owners, managers, workers, and consumers. Fordism designates the overall system of mechanized production, highly rationalized managerial structure, and extensive regulatory control, and usually refers to its fully developed phase in the 1950s and 1960s.[1]

By this time, the state[2] was a much more active mediator and coordinator between capital and society, ensuring steady accumulation, regulating the social costs of growth, and maintaining increased participation of subordinate classes to socioeconomic growth in developed countries. In addition, the system of mass production and mass consumption, with much higher levels of vertical integration, concentration, and rationalization, was also instituted to overcome capitalism's cyclical crises of overproduction and underconsumption. Although sharper distinctions between production workers and managerial, professional, and technical employees (complicated by related racial and gender inequalities) generated new tensions, the labor force was generally pacified by steadily increasing wages, job security, and opportunity for advancement. The overall cooperative relationship between workers and management, or tacit "capital-labor accord," meant that workers in advanced societies and in the primary industrial sector received relatively high wages, improved benefits, and stable employment while planning and regulating the process of production was left entirely in the hands of management. Union struggles were now limited to contesting wages and benefits rather than the organization of the firm and capitalism. This situation was maintained through an intensive exploitation of labor and resources in the Third World, where Fordism signified dependence and underdevelopment[3] (Aglietta 1979; Chandler 1977; Gordon et al. 1982; Harrison and Bluestone 1988; Lipietz 1992; Mingione 1991).

The smooth fusion of capitalism and democracy appeared to come to fruition during high Fordism (1945–1972), a period significantly different from low Fordism (1914–1945). The system also realized conservative hopes for a state-guided mode of capitalism capable of uniting elite interests behind the goal of national development and providing sufficient material advancement to enlist the passive support of the masses. Most importantly, unparalleled levels of abundance moderated the tensions associated with the earlier forms of class inequality. Material and social gains were so great and occurred so quickly that the popular press proclaimed that capitalism's historical contradictions were being eradicated. The Keynesian strategy of sharply increased state expenditures and aggregate demand for goods and services at the local, regional, and national levels helped promote very strong growth. And new regulatory controls promised to keep "externalities" within socially acceptable boundaries. Improved welfare and public education were employed primarily to maintain accumulation and eliminate capitalism's major instabilities without leveling power or redistributing wealth. Class inequality was not diminished, racial and gender divides remained sharp, and corporate hierarchies were strengthened. Yet the system seemed to work well enough for the expanding and increasingly affluent middle class that a desirable alternative to emergent technocracy was hard to imagine. Never before was the legitimation of American-led capitalism stronger, more effective, and consensual (Galbraith 1979; Heilbroner 1993, 149–65; Hodgson 1978, 67–98; Parsons 1971).

Widely popular postindustrial and modernization theories of the later 1950s and 1960s expressed these trends in a much exaggerated manner, implying a fundamental break with the earlier trajectory of capitalist development. The leading American sociological theorist, Talcott Parsons, contended that a noncoercive pattern of "professional authority" was replacing "line-authority" in the United States and in other countries of the advanced West. The Weberian dictum about the "rule of small numbers" was still formally in effect, but the professionally trained labor force was too skilled and specialized to be governed in an arbitrary manner. Bureaucracy was being superseded by "collegial" authority and professional norms of competency, technical efficiency, and responsibility. Because the values of meritorious performance and "instrumental productivism" were now said to hold the upper hand over ascription and private accumulation, the nascent system of corporate power was viewed as a means of implementing collective goals rather than asserting zero-sum interests. The new power and class hierarchies attained a substantive rather than a purely formal-legal legitimacy; the voluntary cooperation of experts replaced compulsory cooperation.

American-led capitalism allegedly constituted a qualitatively new system based on continuous growth, meritocracy, and sociopolitical consensus. The claim was that the important social and economic debates were now restricted to technical and instrumental differences rather than fundamental ideological splits about the overall structure and developmental path of the society. The sudden availability of "the standard consumer package," uplift of subordinate classes into the middle class, and civil rights legislation led social scientists, who earlier portrayed rigid class and status barriers in the 1930s and 1940s, to universalize upward mobility and the middle strata. "Relative deprivation" was supposedly replacing the harsher forms of poverty, and class was increasingly an outcome of differential achievement. As Parsons (1971, 112) stated: "The upper occupational groups . . . far from constituting a 'leisure class' are generally among the most intensely 'working' groups in human history . . . [while] the allegedly 'exploited' working class has moved far closer to becoming the leisure class of modern society." In addition, he viewed access to political offices as wide open to achievement, delimited in power, and responsive to a much more inclusive citizenry. "Functional" and "just" inequality was heralded as the leading principle of social stratification. Overall, the society was becoming a "company of equals" (Davis and Moore 1945; Lipset 1963, 439–56; Parsons 1960, 1964, 1971, 86–91).

Stating explicitly what many others had implied, Parsons (1971, 122–37) spoke of the United States as the "lead society," claiming that a worldwide convergence toward this model was already under way. He and other likeminded Americans equated high Fordism with the realization of the liberating potentialities of modernity; worldwide "development" was an "Americanization" project.[4] The "competing" Soviet style of command socialism, which lacked the formidable fusion of abundance and democracy, was no more than an aberrant detour from the triumphant trajectory of U.S. modernization (Rostow 1960). Against Marxist materialism, postindustrial sociologists stressed the animating force of values in nearly all social action and the American value pattern as the driving force in its economic success. Consequently, cultural modernization, emphasizing the institutionalization of American values, was considered the primary means of generating material and organizational rationalization (Allen 1974; Inkeles and Smith 1974). More importantly, high Fordism was distinguished by its capacity to contain the fragmenting aspects of capitalism within socially acceptable boundaries. It provided a marked, general (though not all-inclusive) increase in the material quality of life, legal and social inclusion, and opportunities for mobility.

Overall, democratization seemed to flow from the new modes of culture and administration.

The Crisis of High Fordism: Economic Restructuring and "Flexibility"

Within advanced capitalist societies, the Fordist regime was challenged by social movements that demanded changes in the redistribution of wealth and in the control of the production process. Political and economic actions led by trade unions were a hallmark of the social terrain of the 1960s. Both in Western Europe and North America, unions were successful in their demands for further increases in worker remuneration and, more important, for the establishment of a greater participatory role in directing enterprises. Increased labor costs translated into declining rates of investment returns for firms. Further, labor's greater control over the production process signified a weakening of the rigid separation between the control and execution aspects of production. In light of these changes, the organization of production in advanced capitalist societies became an "organized negotiation" between labor and management that was mediated by state actions (Dahrendorf 1959; Offe 1985). Union-led movements were paralleled by other socially oriented organized activities, such as students' movements and civil rights movements, which demanded that subordinate segments of the population be granted further participation in all social realms. The political successes of these forces placed constraints on dominant groups' activities. More importantly, they created increased social, political, and economic "costs," which severely taxed the reproductive capacity of the system. In the advanced West, Fordist national systems experienced increasing contradictions centered on their inability to generate enough resources to satisfy both the demands of emerging social groups and those of hegemonic groups.

In addition, new political dynamics arose from the appearance of anti-Western and antimodern movements (e.g., Islamic fundamentalism), resistance to Western modernization models, and new political alignments (e.g., OPEC), which weakened U.S. and other Western countries' positions in international political and economic arenas. These episodes redefined the shape and extent of the flow of wealth from developing to developed countries. Although they were certainly not able to reverse the disadvantaged position of Third World countries, they challenged the existent international division of labor and the terms through which resources in the periphery were made available for the development of the core. By altering the terms of exchange, developing countries established new conditions that forced the developed West to pay higher economic and political "prices."

The oil crisis of 1973–1974, the crisis of the dollar, and the consequent end of the Bretton Woods parity system are some of the embodiments of the realignment process of North-South relations. In essence, firms were confronted with increasing costs of production and diminishing political spaces for action. Governments were faced with declining resources and increasing social demands, resulting in the development of chronic fiscal and legitimation crises. Even the mass media speculated on the possible "end of the American Century," implying that the Fordist regime of accumulation had been terminally eroded (e.g., Bowles and Gintis 1982; Gordon et al. 1982; Harrison and Bluestone 1988; Strobel 1993). Thatcherism and Reaganism signified the deep sense of political and economic crisis and accelerated the movement toward the neoconservative agenda of a strong military and "neo-liberal" (free-market) economics. Under the banner of reduced waste and taxes, the hegemonic neoliberal program attacked Fordist social and economic organization and its mechanisms of social control. In the new political climate, the very organizational hierarchies and socioeconomic policies that were perceived to be the driving forces of Fordist postwar growth began to be viewed as the sources of a crippling "rigidity" that was producing economic crisis and contraction. In arguably one of the most important texts of this period, Daniel Bell (1976, 36) stated:

> Today . . . it is the economic structure that is the more difficult to change. Within the enterprise, the heavy bureaucratic layers reduce flexible adaptation, while union rules inhibit the power of management to control the assignment of jobs. In the society, the economic enterprise is subject to the challenges of various veto groups (e.g., on the location of plants or the use of environment) and subject more and more to regulation by government.[5]

In short, Bell and others maintained that Fordist rigidity placed powerful limits on accumulation. Confronted with declining profits and opportunities for capital accumulation, corporate elites experimented with new strategies for economic growth. They began to search for solutions that would enhance "flexibility" and radically break with the postwar capital-labor accord (Akard 1992). The new antiregulatory sentiments treated state intervention in the economy and society as a destructive weight that undermined accumulation and, consequently, must be dismantled. Although resisted both domestically and internationally, these new strategies matured into that complex of international relations referred to as global post-Fordism.

These trends touched the entire socioeconomic system and within it, the agro-food complex. Lobao (1990, 5) describes the conditions of U.S. agriculture under Fordism as characterized by a "relative prosperity of the farm sector" but also by "increasing scale, decreasing farm numbers and [a] declining middle sector of traditional family farms." She continues by stating that

> Under Fordism, farmers became progressively integrated into circuits of capital, both as producers and consumers. In the first instance, farmers responded to an enormous national market by producing masses of commodities at uniform levels of quality. This entailed integration into output industries, such as food processing, wholesaling, and retailing. In the second, farmers became consumers of mass produced farm inputs from petrochemicals to farm machinery as well as non-farm commodities, notably, processed foods and consumer durables. As a result of these changes, farming became suitably integrated with the industrial system. As long as Fordism was a viable strategy of accumulation, industry and farming could expand and grow. From 1948 to 1972, farm yields rose, government subsidies were manageable, and the cost of food declined as a percent of workers' income. (Lobao 1990, 31)

However, Lobao also agrees that the demise of Fordism beginning in the 1970s eventually affected the organization and structure of the farming sector. For example, she maintains that "midway through the decade, however, serious strains erupted in the wider economy which eventually undermined this prosperity. The postwar, Fordist epoch of high mass consumption, mass production, high wages in family incomes, and liberal welfare-state entered into crisis. By the next decade, rural areas in the farm sector were engulfed by a crisis" (1990, 6).

Global Post-Fordism[6]

Antonio and Bonanno (1995) argue that post-Fordism denotes a restructuring of postwar capitalism that was clearly under way by the middle 1970s.[7] It has involved a concerted effort to diminish rigidity and increase flexibility. Although this tendency includes many multisided processes, operating relatively autonomously in different spheres (i.e., spatial relations, culture, ideology, organization), the most decisive dimension, with the widest consequences, is the elimination of constraints to the free mobility of capital and the maximization of its ease and speed of movement. In the 1980s, the new business-government "partnership" emphasized weakening the community's and state's capacities to regulate its various environments. Although this "economic development" strategy sometimes has been successfully opposed, it has intensified deregulatory pressures throughout the society. Important neoliberal victories have often reduced the voice of the underclasses but also have shifted increased burdens to the middle strata, who generally have supported the policies. Most importantly, post-Fordism has been implemented internationally and is inextricably entwined with the increased "globalization" of capitalism. Following are several important dimensions of global post-Fordism, all related to the strategy of maximizing flexibility.

First, production is decentralized among different owners in various locations. Unlike Fordism where production was unified in vertically integrated firms, often with operations located in a few central locations, the Fordist firm has been decomposed into many subunits and subprocesses carried out by numerous firms spread haphazardly across regional and national boundaries (Harvey 1990, 159, 294–96; Mingione 1991, 198–202). Such decentralization of production should not be mistaken for deconcentration or disorganization of large capital; rather, this flexible strategy enhances corporate control. For example, by divesting certain aspects of the productive process, firms are able to break the bargaining power of unions, transfer risks to other producers, and exploit inexpensive labor or resources of other strategically located firms (Reich 1991; Strobel 1993). It also provides capital with much enhanced leverage in bargaining with the state. Smaller, decentralized operations can choose locations where regulatory and welfare costs are low and organized labor is weak (Mishel and Bernstein 1993; Strobel 1993). Workers understand that higher wage demands or unwillingness to accept cuts will "force" the operation to be relocated. Public bodies operate with the same awareness. Decentralization of production is also used to make concentrated financial holdings more profitable and secure (Sassen 1990). For example, resources from divestments can be reinvested in entirely new forms of production and products, which provide hedges against different possible shifts in the economic environment. By contrast to vertical integration, financial ownership of diverse types of smaller scale production enables firms to act more quickly and efficiently in rapidly changing and fluctuating markets.

Firms use "global sourcing" to seek the least expensive factors of production on an unlimited worldwide basis and to decentralize operations accordingly (Friedmann and McMichael 1989; Gouveia 1994; Sanderson 1986). For example, Sun Valley Thailand is a joint venture between Cargill, Inc., the largest agro-food TNC in the world, and Nippon Meat Packers, the largest meat company in Japan. Sun Valley Thailand sources modern production facilities and the formal production contract system from Cargill, Inc., rapidly expanding consumer markets in Japan from Nippon Meat Packers, and high-quality, low-cost feed, low-paid, docile workers, and lenient government regulations from Thailand (Constance and Heffernan 1991).

Second, even though production is dispersed in many localities, regions, and nations, the financial and research capacity remains firmly concentrated within countries of the First World. Financial global cities (Mingione 1991; Sassen 1990, 1993) orchestrate and control worldwide production, and, similarly, research and development activities remain close by (Busch et al. 1991). Contrary to positions that portray current conditions as chaotic, disorganized, or radically decentered, global post-Fordist flexibility depends on the maintenance of a strong center that is in extremely rationalized control of those major fiscal and intellectual means for exerting command over people and resources spread over wider spaces than ever before. Regardless of the lack of national identity and the tendency to weaken the state's capacity to institute constraints on capital (i.e., rigidities), global post-Fordist firms still require the benefits of legal, social, political, technical, and material infrastructure that is provided by the most advanced and effective state structures. But a primary contradiction of the current situation is that the overarching emphasis on flexibility diminishes recognition of the continued need for coordination and interdependence.

Third, "spatial-temporal" compression facilitates a maximum extension and velocity of economic processes (Harvey 1990). Geographically dispersed, decentralized production combined with concentrated ownership and control requires new forms of instant communication, transport, credit, and other innovative technologies that connect distant operations and rapidly changing locations.[8] The network of transactions is much more complex and depends on extremely multiplicit and flexible informational and financial linkages. Simultaneously, new technology has also accelerated the speed with which material commodities are exchanged. The case of fresh or semiprocessed food is perhaps one of the best examples of increased global spatial-temporal compression (Friedland 1994). The spatio-temporal compression, however, has also involved a reduction in the political blockages, that slow the movement of goods and information. Economic policies oriented toward open markets and corporate strategies to bypass local protectionist policies have also facilitated the global flux of resources (Friedland 1991a; Llambí 1994). It is important to note, however, that the overall increase in speed and mobility of capital has different rhythms (e.g., labor does not move with the same ease as information or finance capital), which contribute to the illusion of disorganization.

Fourth, the spatial-temporal unity of polity and economy, characterizing the earlier phases of capitalist development, has been fractured. Consequently, the state's capacity to mediate between market and society has been weakened. Rather than mediate, the state becomes a more singular agent of unrestricted capitalist development (McMichael and Myhre 1991). This does not mean that all dimensions of the state have been necessarily weakened or that any control function has been eliminated entirely (e.g., police and military power and assistance to capital have often been increased [e.g., Pitelis 1991; Schoonmaker 1993]). The Fordist conception of market-centered democracy presumed the state's capacity to put sociocultural limits on capitalist development and provide community and national institutions with a relative autonomy and safety from the forces of unrestricted economic rationalization. During the later 1970s and 1980s, the state was not able to assure growth and at the same time contain capitalist dynamism without eroding its capacity to limit its socially unacceptable costs.

In particular, global post-Fordism substantially reduced the local, regional, and national state's control over its economic and noneconomic environments (Koc 1994; Ross and Trachte 1990). Post-Fordist firms seek settings with good "business environments." Although this concept often includes qualities such as a skilled labor force and a highly developed and well-maintained infrastructure, it also very frequently means low wages, weak unions, and lax regulation of the workplace and environment. Economic development often means encouraging competitive rollbacks in all these areas (Lambert 1991, 9; Mingione 1991). Moreover, states use tax abatements and various other subsidies to attract or simply keep businesses.[9] In this fashion, post-Fordism seeks external flexibility (i.e., pliable labor forces, publics favorable to deregulation, and cooperative state agencies) in a wide array of interorganizational settings outside the firm.

Fifth, the nature and quality of work is transformed. Full-time employees are being replaced by part-time and temporary workers and manufacturing and farming occupations by service positions. As indicated by Pugliese (1991, 148–49), the nearly universal Fordist model of work has been replaced by a multiplicity of employment arrangements (Bluestone and Harrison 1986; Lobao 1990; Mingione 1991; Newman 1988). Part-time or temporary laborers are hired and released in accordance with changes in market conditions, making around-the-clock operation without overtime pay possible and reducing the costs of benefits and step increases on job ladders. In addition, the workday is reconfigured to enhance flexibility; for example, workers are put on sliding shifts or are "beeped" in during off-hours.

The post-Fordist labor regime has reduced the bargaining power of remaining full-time workers. Older workers are aware that they are unlikely to find equally favorable employment if they lose their current positions. Consequently, they accept "give-backs" in order to maintain job security. The new flexible arrangements surely benefit some workers. For example, many highly skilled and well-paid, high-tech professionals prefer nontraditional schedules in order to work in more relaxed settings (e.g., in their homes), to meet family responsibilities (e.g., dual parenting), or to blend work with other life-style priorities (Leinberger and Tucker 1991, 300–351).[10] Even less advantaged workers sometimes benefit from such flexible arrangements. However, the post-Fordist reshaping of work has meant poorer working conditions, lower wages and benefits, and less job security for a much greater number of workers. Low-wage (and especially minimum-wage) jobs are being created much more rapidly than high-wage ones (Harrison and Bluestone 1988; Newman 1988), and American workers are working longer hours with less vacation time and sick leave. Juliet Schor (1992) contends that the intensification of labor has been so great that many workers cannot meet the demands of other aspects of their everyday life. Lipietz (1992, 105) argues that the postwar capital-labor accord has been restructured into a one-sided vision of the worker as a commodity, who can be freely "borrowed and declined at will by the employer." This costly pliability is at the heart of post-Fordism's coercive flexibility.

Sixth, a new form of transnational capitalism has emerged, qualitatively different from the Fordist type of multinational capitalism (Borrego 1981; Friedland 1991b; Picciotto 1991). Under Fordism, it was possible to identify corporations and products with country of origin (Sassen 1990). In this context, international operations were treated as extensions of entrepreneurial activities designed and engineered in the home country and supported by its state apparatus. By contrast, in the transnational phase, corporate identity and products cannot be connected unambiguously to a particular country (Reich 1991). The automotive industry exemplifies this change. Until a few decades ago, almost all automobiles were built from domestic components and assembled in plants located in the same country. Today, components have diverse national origins and relatively few might be made in the nation where final assemblage takes place.[11] However, companies still market their national identities (e.g., Chrysler's flag-waving "America's back" advertisements). The hypocrisy of the "buy American" slogan is epitomized by Wal-Mart's use of "made in the United States" labels on goods made outside the country. The lack of national identity increases flexibility by reducing loyalty and responsibility to national entities and their economic, social, and political requirements. Maximum flexibility means operating as purely as possible in accord with the bottom line. An example from the agro-food sector again involves Cargill, Inc. Although headquartered in Minnesota, Cargill has a subsidiary called Tradax, which operates from Geneva, Switzerland. Tradax also coordinates Cargill's global agro-food enterprises, which include operations in about fifty countries around the world (Constance and Heffernan 1991).

Finally, cultural postmodernization has accompanied the socioeconomic transformation.[12] Postmodern ideas and sensibilities (stressing sound bites, spectacle, pastiche, simulation, opposition to totalizations, and new identities, modes of expression, and conceptions of space) have had enormous impact in the arts, architecture, media, and cultural interpretation and criticism. Social theorists have strongly emphasized the entwinement of post-Fordism and postmodern cultural fragmentation (i.e., the signifier and the signified lack connection, schizophrenic sensibilities replace the rational subject, pastiche substitutes for style, unified narratives evaporate into a play of differences, and, overall, temporal and spatial confusion reigns) (Bell 1976; Harvey 1990; Jameson 1984). Although the cultural changes should not be reduced to the status of epiphenomena of economic forces, the tendency of global post-Fordism to blur previous sociopolitical, ethnic, associational, economic, and organizational boundaries has had decisive "decentering" effects on cultural representation (Rieff 1993). In particular, spatial borders have been transgressed, compressed, and redrawn in highly nonlinear ways (Gottdiener 1985, 1993; Harvey 1990). The greatly accelerated and highly disjunctive patterns of decentralization, concentration, distintegration, and reintegration produce new modes of economic, mechanical, communicative, and social interdependence that are exceedingly difficult to map and are obscured by the overall sense of fragmentation.[13]

During the height of the Fordist era, in advanced societies the fragmenting aspects of capitalism were maintained within socially acceptable boundaries by the growth of profits, wages, and welfare and by the consequent inclusion of larger and more diverse segments of the populace into the mainstream of middle-class society. Later, the incapacity of the system to maintain stable capital accumulation and the search for flexibility corroded the Fordist social equilibrium. Specifically, flexible accumulation constituted a concerted attack on different facets of the welfare-state mediated capital-labor accord and a rollback to a more inegalitarian form of laissez-faire capitalism. The new strategy of dominant classes restored profitability at the cost of a substantial weakening of the state's welfare, redistributive, and regulatory powers, sharply increased class inequality, and produced intense sociocultural fragmentation. Now capital accumulation reversed the trend toward an inclusive middle-class society, decoupling the "unity" between capitalism and the postwar progress toward the establishment of patterns of increasing living standards.

The increased class polarization and a smaller, more insecure middle class is made evident by data on the socioeconomic conditions of many countries of the developed West. In the United States, for example, family income steadily increased from 1947 until 1973 at an annual average rate of 2.3 percent per year or a total of $12,706 for the entire period. This growth slowed to a rate of 0.6 percent after 1973 and further declined to 0.4 percent, or $153 per year, in the 1979–1989 period. From 1989 to 1991, the average family income decreased by 4.4 percent, despite increases in two-income families (in the longer period between 1978 and 1989, "female spouses" worked 32.2 percent more hours on average), productivity (up 1.3 percent from 1970 to 1979 and 0.8 percent from 1979 to 1989), and overall jobs (6 million more jobs in 1990). In the last decade, family income of the top 1 percent of the population grew by an unprecedented 62.9 percent, capturing 53.2 percent of the total family income. This figure was up from 35 percent of the total family income in 1959. By contrast, the income of the bottom 60 percent of the population declined (–1.0 percent). From 1989 to 1991 the bottom 80 percent lost income (2.3 percent), while only the top 20 percent gained. Workers with some college training in 1989 were paid 8.3 percent less than in 1979, workers with a high school diploma experienced a 12.7 percent decline in wages, and workers with no high school degrees were paid 18.3 percent less. Young workers (25 years old or younger) experienced perhaps the worst decline, since they are paid 22.4 percent less than a decade ago (Bartlett and Steele 1992, 140; Mishel and Bernstein 1993, 2–4, 14, 16, 19–20, 33–41, 43–45, 49–54). In short, the working class in the United States as well as in other advanced countries is laboring harder yet is being remunerated significantly less in regard to its output. Nearly two decades of shrinking real wages and benefits have undermined the material well-being of a growing portion of the middle and working classes and have generated a much more widespread "fear of falling."[14]

As Mingione (1991) has argued, current conditions amount to a "fragmented polarization" linking growing class disparities to divisive work conditions. By contrast, Fordism had a much greater uniformity of production roles, tended to concentrate many workers under one roof, and even unified spatially dispersed workers under one ownership in giant vertically integrated corporations producing common product lines. Although labor-force segmentation had increased substantially from the early days of Fordism, the characteristic corporate firm of the postwar era still provided temporal-spatial unities that favored worker association, communication, and organization. The cooperative nature of capital-labor relations during the Fordist period reflected the leverage that these unities afforded the working class. By disempowering labor through the fragmentation of work, post-Fordist flexible accumulation strategies broke the capital-labor accord, which was based upon the achieved strength of labor in the Fordist era.

The highly spatially dispersed and variegated labor forces of the smaller and less stable post-Fordist industrial firms have undermined labor solidarity and facilitated political efforts to weaken unions during the 1980s (Gordon et al. 1982; Harrison and Bluestone 1983; Harvey 1990). Under post-Fordist fragmentation, workers identify with their geographic location of residence, ethnic background, nationality, and consumption patterns rather than with other members of their class. Indeed, post-Fordist capital mobility and economic development pits workers against their counterparts in other areas of the nation or world.

The Debate on the Transition

James O'Connor's Classificatory Analysis

The transition from Fordism to global post-Fordism has been analyzed by a number of scholars. O'Connor (1989, 21) divides the explanations for the various aspects of the current restructuring of capitalism into four groups: bourgeois economists using market theory, neoorthodox Marxists using value theory, and neo- and post-Marxist theorists using social theory and social psychological theory, respectively. Bourgeois economists concentrate on the integration of capitalist economy at the level of exchange or market relationships. Neoorthodox Marxist value theorists study capitalist system integration at the level of production and circulation of capital and capital accumulation. Neo-Marxist social theorists focus on social integration in capitalist society, and the subject of the post-Marxist personality theory inquiries is personality integration. According to O'Connor, (1) none of these approaches is necessarily better than the other, (2) each is a successively more concrete, less deterministic, and historically contexted interpretation than the one proceeding it, and (3) each is an increasingly less partial and more substantive view of reality (1989).

O'Connor's agenda is to advance a critique of economic determinism and highlight the explanatory power of the neo- and post-Marxist approaches, which focus on the social and social psychological realm of material life. The current restructuring phase of capitalism should not be interpreted as "merely an objective historical process" but as "a subjective historical process—a time when it is not possible to take for granted 'normal' economic, social, and other relationships; a time for decision; and a time when what individuals actually do counts for something" (1989, 21). O'Connor argues for "opening up crisis theory to more interpretive and less deterministic approaches, meanwhile attempting to avoid simple subjectivism and voluntarism" (1989, 22).

According to O'Connor, the theory of market equilibrium based on the totality of exchange relationships is the most formal and objectified approach to crisis theory. The market via the universal medium of money is the process by which all commodities are exchanged. Real individuals become the "bearers of commodities" and thereby "objectify themselves and others" (1989, 22). Crisis in equilibrium occurs when the transfer of property rights between capitalist enterprises and individuals breaks down. From the bourgeois worldview of market theory, the current restructuring characterized by inflation, unemployment, large state deficits, and Third World debt is the result of (1) market disequilibria and/or inadequate monetary incentives and penalties and/or (2) disparities between physical production capacity and effective demand for commodities combined with excessively high interest rates. Disequilibria and disparities are the result of excessive government, labor unions, or other interference with free competition. For O'Connor, a good example of the views of bourgeois market theorists Hobson, Schumpeter, and Veblen is "that the current problems in capitalism are in large part the result of excessive government regulation and public deficits and borrowing which raise interest rates and hence discourage investment spending" (1989, 23).

The neoorthodox Marxist value theory of the current transition from Fordism to global post-Fordism is based on social labor and social class relations—a more concrete level of capitalist society than market theory. The central issues are (1) the theory of labor exploitation and its relationship to economic crisis focusing on the extraction of relative surplus value and (2) the theory of the contradiction between productive forces and production relationships. Value theory, like market theory, retains the convention of abstracting from social-cultural life as well as real individuals. For neoorthodox value theory Marxists, deficits, unemployment, and inflation are the result of inherent contradictions of capitalist relations and production, such as class struggle and trends toward concentration of capital. These contradictions make crisis inevitable and are indeed necessary to recast the relations of production in more efficient arrangements. Although economic crisis is inevitable, it may be postponed by imperialist expansion, the growth of credit money, state monetary, fiscal, and Keynesian welfare policies, and the restructuring of physical production and social relationships between labor and capital. For this camp, the current crisis is in large part due to the result of capital overproduction and the tendency of the average rate of profit to fall. The crisis is postponed through the expansion of credit money and fictitious profits, which displace the crisis into the political sphere in the form of inflation or massive Third World debt. Although orthodox neo-Marxists have a powerful method for investigating the global movements of capital commodities, by regarding individuals and groups as personifications of capitalist production forces and relations their approach, much like the bourgeois economists, lacks the ability to grasp the social and political meanings of these global movements, thereby negating an understanding of the "real possibilities of social struggle and social movements" (O'Connor 1989, 25).

Neo-Marxist theories of social crisis are centered on the concept that "real" individuals in historical time and space are implicitly and explicitly defined as (1) personifications of social categories such as youth, women, and the elderly, or racial minorities and (2) quasi-groups such as environmentalists or feminists. By incorporating market and value theories with historically centered groups of individuals, neo-Marxist social theorists "come closer to representing the 'concrete totality' [in Karl Kosik's words] of the social, cultural, ideological, and political economic contradictions which constitute the modern crisis" (O'Connor 1989, 27). From this point of view, unemployment, inflation, deficits, and the other manifestations of the current crisis are the result of contradictions within and between political and social structures within and against developed capitalist social and ideological structures based not only on labor issues and class struggle but also on other identities such as nationality, ethnicity, or gender. Social theorists reject the notion of an economic determinism that carries with it a rejection of economic crisis theory; instead, they advance a theory of social and political crisis struggle and crisis that "are and are not part and parcel of economic crisis and economic struggle" (O'Connor 1989, 27). The neo-Marxist's critique of the dualism between objectivist and subjectivist approaches is based on the idea that modern economic, political, and cultural crises interpenetrate each other in ways that transform them into different dimensions of the same historical process—the disintegration and reintegration of the modern world. In other words, the working class, social and cultural movements, the state, and society as a whole are themselves implicated in the development of the forms and contents of the modern crisis (O'Connor 1986).

Personality crisis theorists focus on the premise that the most concrete level of social life is the day-to-day lived experiences of real, social individuals, individuals who at the same time occupy positions in capitalist relations and are irreducibly unique persons. Personality approaches include forms of psychological repression, subliminiation, and projection that guard against individual self-knowledge and also views and experiences of capitalist alienation, exploitation, and reification that include "the social and cultural processes of the production and evolution of experience of meaning as well as inner personality conflicts and struggles" (O'Connor 1989, 28). Personality crisis theory is (1) a critique of the self-deceptions that individuals use to legitimate themselves as self-defined moral beings, (2) an explanation of the value of modern repression, such as distorted communication that facilitates successful social integration into capitalist structures, (3) an account of individual struggles against ideological, individualistic identities and for a richer social identity, and (4) the relationship between struggles for social individuality and the social construction of crisis as struggles within and against the capitalist state and political apparatus. In summary, political struggle and its relationship to personality struggle are "defined not only in terms of means to particular material or social ends but also as ends in and of themselves" (O'Connor 1989, 29).

Following Jessop (1990), we illustrate the debate among European scholars. Particularly, reviews of the Parisian regulationist school and of the British school are presented.

The Parisian Regulationist School

The Parisian Regulationists[15] began their research program in discussion of Aglietta's original works on inflation (1974, 1975), which were concerned with Fordism in the United States, the nature of monopoly capitalism, the causes of inflation, and the development of public spending by the state. In contrast to the Grenoblois and orthodox state monopoly capitalist theories, the Parisian school distinguishes only two basic stages of capitalism: extensive and intensive. In an extensive regime of accumulation, capitalism expands mainly by spreading into new areas of activity at the expense of precapitalist forms of production. In an intensive regime, capital is accumulated primarily by reorganizing existing areas of capitalist activity, especially the labor process, in order to increase the relative surplus value. This school claims that an extensive accumulation regime is associated with a competitive mode of regulation and an intensive regime with a monopoly mode. The modes of regulation are defined in terms of the wage relation. Competitive regulation involves a flexible wage formation, while monopoly regulation is characterized by collective bargaining mediated by the welfare-state and rising consumption norms. Competitive regulation is based on metallic money while the monopoly mode is based on credit and state money. Parisian regulationists have looked mainly at the development and dynamics of Fordism, neo-Fordism, and post-Fordism considered as "regimes of accumulation" and/or "modes of regulation" in specific national economies and stress the heterogeneity of their national variants (Aglietta 1978, 1979, 1982, 1986, 1988; Aglietta and Brender 1984; Boyer 1979, 1985; Boyer and Bowles 1988; Boyer and Mistral 1984; Lipietz 1986, 1987b, 1992). In their more recent works the Parisian theorists have distinguished more regimes of accumulation and discuss transitional periods in greater detail (Jessop 1990).

Aglietta's *A Theory of Capitalist Regulation* (1979) is the anchor of the Parisian regulationist school and a polemic against ahistorical classical economics. Aglietta's work (1986, 1979) and Lipietz's work (1987a, 1987b, 1991, 1992) provide the underpinnings for the current uses of regulation theory. Aglietta argues that his concepts of the "regime of accumulation" and associated "mode of regulation" represent an epistemological advance for Marxist analyses as an important mediation between the theoretical insights of an abstract form of social relations based on the law of value and the concrete empirical nature of reality as revealed using the methods of historical materialism. The historical evolution of the wage relation is the focus of the regulationist school. Aglietta analyzes the evolution of capitalist accumulation from extensive to intensive forms conceptualized as expressions of absolute and relative surplus value. These forms are always historically and geographically specific in relation to particular national territories.

Concerning the case of the United States, the Parisian regulation school sees the latter part of the nineteeth century and the beginning decades of the twentieth century as a period of extensive accumulation strategies associated with a wage relation based on absolute surplus value extraction and combined with a competitive mode of regulation. After World War I and the Great Depression, the U.S. economy shifted to a predominantly intensive mode of capitalist accumulation based on relative surplus value creation combined with a monopoly mode of regulation. The passage from one regime of accumulation to another produces a profound and enduring change in the capitalist social relations, with particular implications for the production and reproduction of the wage relation.

From the regulationist view, access to labor in capitalist society is a condition of life that goes far beyond the immediate relations of the exchange economy and depends on a historically specific set of social norms that are institutionalized within civil society by the state. These social norms make up the "mode of regulation." The regularity provided by the mode of regulation is always relative, and its precise terms are constantly being reconstituted by class struggle.

The particular character of the modes of regulation in the forms of institutionalized social norms is influenced by two factors: the historical development and character of the nation-state and the specifically national character of the social development of capitalist reproduction with particular nation-states. The social development of capitalist reproduction and the evolution of the wage relation in the United States is divided into three periods, with each regime of accumulation presenting as part of its mode of regulation its particular set of social norms on the wage relation. Aglietta refers to the period of the mid-1800s as the extensive regime of accumulation based on the "frontier principle." The frontier principle gave way to the Fordist intensive regime of accumulation between the two world wars, and the Fordist era lasted until the 1960s when the transition to post-Fordism began.

The frontier principle and the extensive regime of accumulation are characterized by the social and territorial expansion of wage relations. During this period social relations were predominantly based on the pursuit of absolute surplus value, i.e., hiring more workers and working them longer hours (Lipietz 1987b). Absolute surplus value is defined as the effect of the wage relation as a process of separation between individuals and also the separation from individuals of the means of producing their own existence (Aglietta 1979, 71). The frontier principle includes a rural transformation process based on the consolidation of production within the bounds of a commodity economy where agriculture becomes increasingly subordinated to the development of an industrial foodstuffs complex. Political struggles in this period tended to revolve around the institutionalization of the nation-state and the social norms involved in the utilization of labor-power in production, including the length of the working day, the determination of wage levels, and the beginnings of the "socialized management of the reproduction costs of the wage earning class" (Aglietta 1979, 74). This is the period when society is homogenized spatially, and individuals become undifferentiated, interchangeable components of production and consumption.

From the regulationist view, the extensive regime of accumulation was characterized by (1) the creation of an independent national territory with sovereignty over a vast geographical region (a nation-state), (2) the political and juridical struggles over the institutional necessities for absolute surplus-value production such as factory and labor regulations, and (3) the problem of coordinating rural production within industrial commodity production. The particular American character of the frontier principle included the achievement of homogeneity through a colonial insurrection and a later bloody civil war, the imposition of new social norms facilitated by the absence of ties with precapitalist forms of production, and the vast availability of land and mineral resources as an extensive means of surplus value creation. In all three cases the westward territorial expansion of wage-based commodity relations was extended and codified.

The post–Civil War recessions in the 1870s began a period of laissez-faire economic philosophy that included the belief in limited government, equal opportunity, and the benefits of social "natural" selection, all aspects of social Darwinism. This period was characterized by the consolidation of heavy-industry capital goods and bloody struggles over labor legislation and the problems of the proletariat. The battles between capital and labor provoked major political disputes over the proper role of the state in relation to the emerging social order.

The Great Depression represented a historical moment in the transition to an intensive regime of capital accumulation. The development of Taylorist scientific labor management provided for a new regime of accumulation based on the intensive production and appropriation of relative surplus value. Mass-production technologies expanded rapidly in the early 1900s and increased production beyond the ability to consume. Productivity grew at rate of 6 percent per year (three times the average rate of the nineteenth century), but the purchasing power of the workers "remained mediocre" (Lipietz 1987a, 28). The Great Depression was a "typical over-production crisis" (Lipietz 1982, 110). A new regime of accumulation had emerged based on Taylorist mass production but the complementary mode of regulation was not yet institutionalized. The New Deal with its "class compromise" provided the needed monopoly mode of regulation.

The United States emerged from World War II with an intensive regime of accumulation and a monopolistic mode of regulation in place based on the balancing of mass production and mass consumption—the Fordist regime. The social norms and institutions (mode of regulation) that connect mass production and mass consumption were social security provisions, collective bargaining, contractualization of labor, real wage increases tied to productivity gains, oligopolization or direct state regulation of major markets, credit money as legal tender, and Keynesian economic policies (Lipietz 1984, 86). Fordism incorporated mass consumption into the accumulation process via a politically regulated linking of wage, price, and productivity (Lipietz 1987b, 35). Under Fordism the workers were paid enough to buy the durable goods they produced. The growth of automobile and home ownership on credit exemplified the solidifying norm of mass consumption.

Workers in the frontier era often were recent immigrants and lived in the huge ethnic neighborhoods in the large cities. They often walked to work. Workers in the Fordist era more typically drove cars from suburban neighborhoods to work. Fordism is named after Henry Ford, who argued early on that workers needed to be paid enough to buy what they produce. He paid his workers $5 per day—enough to make payments on a Model T. State-regulated systems of consumer credit combined with rising wages helped produce the social norm of consumption characteristic of the monopoly mode of regulation. The rise in consumer purchases to furnish new homes and changes in travel distances involving work were central to the new regime of intensive accumulation. Fordism was as much a way of life as it was a production system. Mass production meant huge lots of relatively undifferentiated products that would be bought by a mass of relatively undifferentiated consumers. Aglietta summarizes this passage to Fordism as a "long historical process which began at the start of the 20th century, and has seen the penetration of capitalist production into the internal organization of towns, and into the production of the means of individual consumption for the broad mass of wage-workers" (1979, 74).

The new result was a rapid rise in the rate of surplus value creation through not only Taylorist-Fordist reorganization of production but also the transformation of the reproduction of labor power. The effectiveness of labor power was to be substantially raised by concentrating rational operations in the factory and relocating the reproductive functions to the decentralized residential neighborhoods. This would also be the material base for the new social norms being institutionalized in this regime. The "social norm of consumption" required the adaptation of use value characteristics of the objects of consumption to mass production (Aglietta 1979, 82). The structuring of this norm also included the socialization of expenditures by wage earners through credit. The rate of penetration of the new commodities aimed at labor reproduction became a function of income differentiation. The pursuit of relative surplus value in the intensive regime structured the wage-earning class in a spectrum of required differentiations which segmented the labor market.

After the reconstruction of Europe and the Korean War, the Organization for Economic Co-operation and Development (OECD) countries experienced Fordism's "golden age." For about fifteen years they enjoyed an exceptionally strong and regular growth, and consumption gains were matched to productivity gains "more or less explicitly by a policy of regulation of the wage relation" (Lipietz 1987a, 29). The characteristics of the institutional forms that made up the Fordist mode of regulation included (Lipietz 1987a, 29):

1. A collective "contractualization" of the direct wage. Most workers worked under collective agreements, and a minimum wage was established. Employed workers would be able to consume at least at minimum levels.

2. The welfare-state provided a social security system plus an unemployment system. These programs also assured continued cash consumption from the elderly and laid-off workers.
3. The growth of the service sector of management, trade, and finance.
4. Important modifications in the relations between banks and industrial firms that allow firms to redirect their production from division to division while maintaining their prices in the declining divisions.
5. The role of the state was modified to include the management of the wage relation and the management of the currency. The management of the currency became the backbone of Keynesian economic policy.

Fordism was based on the growth of the national internal market and on matching domestic production to domestic consumption; international trade to external markets was secondary to Fordist accumulation. The South played an important role as the provider of cheap labor and raw materials. Continued access to these materials was the "essential task of the political and military hegemony of the United States" (Lipietz 1987a, 30).

International trade did develop between OECD countries. Lipietz calls this the "catching up" movement from Europe and Japan. After World War II the United States "imposed its model of development, culturally at first, then financially (with the Marshall and MacArthur plans), and finally institutionally (Bretton Woods agreements, the creation of the General Agreement on Trade and Tariffs [GATT], the International Monetary Fund [IMF], and the OECD)" (Lipietz 1987a, 31). Under these arrangements, no form of international regulation of the wage relation was necessary. Across Europe and Japan the same principles of "contractualized collective bargaining," "the welfare-State supported minimum wage, social security, and unemployment insurance," and "rising wages and credit supports supported sustained purchasing power" (Lipietz 1987a, 31).

The strength of the U.S. economy and dollar backed the production of goods consumed both domestically and internationally. To increase consumption in Europe and Japan, the U.S. state extended loans and U.S. MNCs increased their direct investment. These loans and foreign direct-investment dollars theoretically were backed by U.S. gold reserves. By the time France demanded gold for its euro/xeno dollars and the United States refused, the undisputed domination of U.S. capitalist production was a thing of the past. According to Lipietz, the system of international trade

> is not, therefore, a regime of accumulation, strictly speaking, but a world configuration which provisionally realizes the compatibility of a juxtaposition of similar accumulation regimes, differentiated by individual rates of growth of modes of international insertion. Schematically, the United States equips Europe (or Japan) in exchange for a claim on European labour power, the purchase of this labour power (by multinational firms) in turn giving rights to U.S. capital goods, the purchase of which, linked to the accelerated generalization of Fordism, permits Europe and Japan to catch up. This catching up does not occur in the "import substitution" countries of the South, because they lack sufficient social reforms.
>
> The world economy could never surpass this level of implicit organization. There could never be institutional forms regulating world demand; and there will never be any supra-national sovereign authority regulating the money supply. The complementarities and antagonisms of the national economies would always remain "configurations" of relationships, partial, and unstable, and it is only as a figure of speech that it is possible to talk about any "world regime of accumulation." (1987a, 32)

To add to Lipietz's assertions, Aglietta argues, "It cannot be overemphasized that there is no universal mechanism of international adjustment on which a remodeled hierarchy of nations can be simply grafted" (1982, 14). International regimes of accumulation and modes of regulation are much more problematic than national arrangements.

Aglietta (1979) argues that by the late 1960s the Fordist era entered into crisis based on declining productivity, rising wage costs, and market saturation. The "blue-collar blues" of the late 1960s represented the exhaustion of the Taylorist-Fordist production paradigm (1979, 119–21). As economic growth slowed globally, the postwar capital-labor accord unraveled and the United States experienced a balance-of-payments crisis. The Bretton Woods agreements collapsed as the dollar was taken off the gold standard, and flexible exchange rates ensued. The world economy and its constituent First World economies entered into an organizational crisis that was accelerated by the growth of unregulated global financial institutions such as the Eurodollar markets. As nation-states lost their ability to regulate their national monies, massive international movements of money produced economic destabilization (Aglietta 1979). "This became the condition for a global accumulation process which restructured production, markets and class relations on a world-wide basis" (Clarke 1988, 349).

Lipietz presents two general explanations for the crisis of Fordism: first, the contradiction between national monopoly regulation and internationalization of production; and second, the exhaustion of productivity increases associated with Taylorist/Fordist production principles. Productivity gains began to slow at the end of the 1960s resulting in an increase in the value of capital per head or the organic composition of capital. Lower productivity led to reduced investment, which led to rising unemployment, which led to increasing costs for the welfare-state in the form of income supports (1987a, 34).

As opposed to the 1930s crisis of overproduction, which French regulationists attribute to the emergence of the intensive regime based on Taylorist mass production without an appropriate mode of regulation to support consumption, the current crisis is one of profitability. The welfare-states' assorted institutional forms of monopolistic regulation limit the recessional effects of the fall in productivity through unemployment insurance and credit availability, thereby moderating the effects of the crisis into inflation and stagflation instead of depression.

Lipietz argues that the cause of the slowing in productivity gains was due to exhaustion of Taylorist/Fordist production efficiencies combined with the resurgence of working class struggles in the late 1960s and early 1970s. Fordism had finally found the "one best way" to produce certain commodities and a "counter-offensive of specialized workers" rose up to avenge "the defeat of the professional worker" (1987a, 37). These factors put the regime of accumulation in a vulnerable position.

By 1967, the productivity of Japan and Europe (West Germany and France in particular) approached that of the United States. The difference in wage costs per unit produced became unfavorable to U.S. competitiveness. The rate of investment in both Japan and Europe was higher than in the United States. The growth of MNCs using Fordist methods in Europe provided the fuel for the "catch up." From 1967 onward the U.S.

trade balance was in deficit. Due to that loss of U.S. economic competitiveness, the dollar became overvalued. With the demise of Bretton Woods, the dollar was no longer a fixed international currency but rather floated on international markets. Foreign subsidiaries of MNCs began to finance themselves directly from external sources. A trade war emerged between the tripolar economies (the United States, Europe, and Japan) "in which the phases of expansion and recession tended directly to echo and amplify one another" (Lipietz 1987a, 38).

The boom world of 1973 was followed by recession in 1974. At the same time, U.S. hegemony in international affairs was rapidly eroding. The U.S. failure in Indochina provoked a wave of nationalism in the Third World, most notably the Arab-Israeli War and resulting emergence of OPEC, which "gave the petroleum exporting countries the opportunity to recover control of the fixing of oil prices" (Lipietz 1987a, 39). In developing countries, the general crisis of Fordism created inflation, which pushed firms and governments to squeeze wages and restrict credit. These actions resulted in further decreased global demand and consumption. Again, the welfare-state prevented a "depressive spiral" through the use of unemployment insurance, which guaranteed at least minimal levels of consumption. As the 1970s progressed, Keynesianism frequently had to come to Fordism's rescue.

The Fordist nations used "social-democratic" strategies to deal with the crisis. Carter was the U.S. president, social-democrats were in power in Germany and Scandinavia, the Labour Party was in power in England, and in France Giscard d'Estaing and Chirac organized the recovery. The nations "absorbed" the oil shock by either borrowing or printing more money. Carter supplied the world with a credit currency of dollars that provided millions of jobs but also devalued the dollar, making U.S. production and goods more competitive internationally. Japan used this period to capture market shares and flood the market with Fordist-type manufactured products. The European economies experienced weaker growth during this period due to the "great rigidity (in terms of social relations) of the European variant of Fordism" (Lipietz 1987a, 44).

Even though Europe is more and more integrated at the industrial level, it remains fragmented into distinct nations that must each manage their trade balance. In order to maintain a trade balance, a country must import less, consume less, and invest less than its trade partners. Accordingly, the wage cost becomes one of the determinants of international competitiveness. Some form of "collective wage" and "minimum wage" is necessary to keep firms from competing by decreasing wages and therefore depressing demand. Like unregulated firms, the European countries found themselves in a situation of "competitive regulation" strategies to balance trade. The factors such as collective agreements and minimum wages that were necessary for an economic truce and accumulation were lacking in Europe. In other words, because these kinds of agreements are necessary for a regime of accumulation to function, they have so far been obtainable only within a nation-state context. To the present, "there are no international collective agreements, there is not even a European social space" (Lipietz 1982a, 44).

Even though there are no international wage agreements that could provide some basis for an international regime of accumulation, the circulation of OPEC dollars, TNC foreign direct investment, and U.S. deficit spending to stall the recessionary aspects of the crisis provided for Fordist investments and increased consumption in the emerging NICs a form of "world Keynesianism" that fostered the growth of peripheral Fordism. For Aglietta (1982), it was the development of this international debt economy based on money markets beyond the regulation of any nation-state that destabilized the "virtuous circles" that characterized the Fordist era and introduced the "vicious circles" of stagflation.

Peripheral Fordism was another one of the ways that capitalism extended the life of Fordism. Peripheral Fordism is the "transfer of vulgarized Fordist production to low wage and low social security countries" (Lipietz 1987a, 46). The triumph of the NIC in the 1970s essentially consisted of the adoption of the Fordist labor process without the complementary regulation of the wage relation. The success of NICs was based on importing the North's raw materials and then selling manufactured products back to the North. NICs financed their imports on credit, which expanded the international credit currency (xeno dollars) based on U.S. credit currency, and used this credit from the surplus nations (OPEC and Japan) to purchase the capital goods from the OECD. NICs then sold the manufactured goods back to OPEC and OECD and used these monies to pay back the loans. During the 1970s this strategy worked well for many countries because (1) the productivity gains of NICs were made very rapidly and the hourly wage rate was five to ten times less than in the North, and (2) the world market had continued to expand because of the national and international Keynesian management of the crisis (Lipietz 1987a, 44–45).

For a few years many NICs were able to attain rates of growth exceeding 10 percent. Lipietz argues that the drive for import substitution and the resulting embrace of peripheral Fordism has a similar effect on the South as the Marshall Plan had on Europe and Japan. As OPEC oil dollars were loaned to NICs to industrialize, the Fordist labor process was also installed. "Extremely dependent on the evolution of world demand, it affected only a very limited number of countries of the South, at the same time destabilizing their internal structures" (1987a, 45).

Because of the welfare-state's Keynesian policies, Fordism survived the 1970s, but its fundamental crisis was becoming more and more evident. The combination of "abnormally" slow growth and an increasing rapid "inflation" produced "stagflation." Peripheral Fordism was one of capitalism's strategies to escape the crisis, the major strategy being a reconstruction of the wage relation. Historically the productivity–direct wage link between the worker and the firm provided the main support for labor power reproduction; the welfare-state supports were secondary. As the role of the welfare-state was expanded, especially during the Great Society, to where it virtually insured a permanent income to wage earners and their families, the firms attempted to free themselves from their contractual obligations to their employees, such as wage increases and fringe benefits. The permanent workers of the "core" began to dwindle and be replaced by an army of temporary or part-time workers, who survived on welfare handouts and infrequent wages. As capital broke the accord with labor, "the last traces of 'status' or of a 'trade' were disappearing for young people, condemned from their entry into the labour force to a succession of 'small jobs' and welfare" (Lipietz 1987a, 46).

Another strategy that came out of the crisis was the search for new sources of labor productivity. It consisted of both the promise of an electronics-based technological revolution and a replacement of Taylorist principles of separation of tasks with a reaggregation of tasks and the involvement of "quality-control groups." This combination produced spectacular results in Japan in the late 1970s and 1980s. Indeed, "Japanese capitalism did not 'catch up' with the United States. It overtook it, by inventing a new, post-Fordist method of transforming the manual and intellectual producers' ingenuity into productivity" (Lipietz 1987a, 47).

Through the late 1970s industrial production in the OECD countries improved steadily. The "$11 per barrel" first oil shock of 1974 had been absorbed by world inflation; the "$34 per barrel" second oil shock of

1979 was handled in a different manner. According to Lipietz, "Everything was to turn out, in fact, very differently, as if the dominant classes no longer believed, or could no longer believe, in Keynesianism" (Lipietz 1987a, 47). The second oil shock led to the monetarist shock of 1981.

Keynesian policies were losing their effectiveness. Trade deficits in France and Great Britain and large public debts in the United States, West Germany, and Japan reduced the options of the "Keynesian game." Domestic markets were saturated, and First World industrial economies began to penetrate each other's economies through TNC mergers and acquisitions of firms previously competing with each other. More importantly, the world elites, the heads of firms and politicians of the Trilateral Commission, stopped believing in international Keynesianism (Lipietz 1987a). By the end of the 1970s, growth remained slow, fixed capital per capita continued its acceleration, and U.S. currency continued to lose its international purchasing power. The victory of Thatcher in England and Reagan in the United States, the appointment of Volker as head of the U.S. Federal Reserve Bank, the victory of the center-left coalition in Germany, and the full implementation of Barrism in France after the defeat of the left in 1978 all reflected the move from Keynesianism to monetarism. The ideas of liberalism and a trust in the market were reasserted in propositions such as "Let the market reward the firms that can vision the future and eliminate those that are slag of the past" (Lipietz 1987a, 48).

When Ronald Reagan became president, the extreme monetarist camp of Great Britain and the United States was created. Both administrations cut welfare spending and credit support with the result that "these two coalitions erased in their respective countries within a few months the growth of the previous five years" (Lipietz 1987a, 49). The recession they provoked dried up the creation of international credit by erasing OPEC surpluses. Credit became costly and scarce. World demand quickly decreased and the NICs' strategy of peripheral Fordism found itself unable to make debt payments, which resulted in bankruptcies. By 1983 a world financial crisis loomed ahead, and the United States relaxed monetary constraints. When Mexico declared a suspension of its debt payments, the Reagan administration abandoned its monetarist strategy, ordered the world banking system to accept a general restructuring of Third World debt, and also restored domestic Keynesian policies. Tax cuts and increased defense spending pumped up the economy and produced a giant deficit largely financed by the Federal Reserve. Housing recoveries led the U.S. and German economies and exports led the Japanese recovery. Still, many NICs could not repay their debt, which hindered the "multiplying mechanisms" of the xeno-dollar creation that had kept Fordism on life-support systems (Lipietz 1987a, 50).[16]

The British School

The British regulationists Lash and Urry (1987) focus on the concepts of "organized" and "disorganized" capitalism. The central theme of their analysis is that capitalist countries are moving from an "organized" to a "disorganized" structure under the pressure of the present postmodern transformations in time, space, economy, and culture. Through historical analysis of the social, economic, political, cultural, and spatial developments in five case studies (Germany, Sweden, Britain, France, and the United States), the authors trace the transition from "liberal capitalism" to "organized capitalism" around the turn of the century, through its progressive breakdown since the 1970s to the emergence of a new phase, "disorganized capitalism." Organized capitalism, which is what Marxists and Weberian social scientists have typically analyzed through such concepts as finance capitalism, bureaucratic capitalism, state monopoly capitalism, increasing centralization, and rationalization, is coming to an end, and the era of disorganized capitalism is beginning.

The authors' notion of organized capitalism was developed from the works of Hilferding and Kocka and begins in the "final decades of the nineteenth century as a consequence of the downward phase of the Kondratieff long waves which began in the mid-1870s" (Lash and Urry 1987, 3). During this period the combination of a worldwide recession and Britain's dominance in international markets "engendered the shift of its competitors into organized capitalist development" (1987, 114).

Organized capitalism is characterized by the progressive concentration, coordination, and integration of industrial, financial and commercial capital that undermined liberal competitive or free-market capitalism. Companies became increasingly bureaucratized as evidenced by the divorce between ownership and management and the institutionalization of internal planning departments. The rise of state interventionism in the form of regulation, national planning, and the welfare-state transformed politics and gradually obliterated the boundaries between state and civil society. The expansion of state apparati was matched by a similar bureaucratization of civil society through progressive organization of labor into national, bureaucratized unions that were then incorporated into state structures through corporatist or neocorporatist class compromises. An ideological commitment to modernism, scientific rationality, and technocracy underpinned the solidification of "organized capitalism."

Organized capitalism's development is a national phenomenon, linked to the rise of nationalism and delimited by national boundaries. Accordingly, the path by which organized capitalism was installed in each case and its eventual configuration varied according to the structural particularities of the five countries. The ideal type of the system was Germany, which from the 1870s was organized at both the top—through the concentration and cartelization of industry and the interarticulation of banks, industry, and the state—and at the bottom—through national labor organizations, class-based politics, and the welfare-state. The other cases differed in both the timing and extent of organization. Sweden was organized from the bottom and not until the interwar period. France was organized very late and only through the intervention of the state. America was organized very early but almost exclusively at the top. Britain organized very early at the bottom, but only very late and weakly at the top. These variations on the common trajectory toward organized capitalism are attributable to a nation's size, the timing of its organization, and the degree to which precapitalist structures persist. Despite the widely divergent patterns of organization, the eventual systems were remarkably consistent. For these authors, there is an underlying historical imperative that both required organization and gave it a common appearance in widely different national environments.

Their theory is as follows: The earlier countries develop, the less organized capitalism will be due to the high cost of late organization; the more precapitalist organization that survives into the capitalist period, the greater the degree of organization; smaller countries had to become more organized in order for their industries to compete internationally; and the less organized capitalism becomes, the more quickly it becomes disorganized. On this basis, Lash and Urry account for the distinct patterns of capitalist organization in the five countries.

If organized capitalism is a national phenomena, disorganized capitalism is global. Transnational corporations in industry, finance, and commerce—which straddle not only national borders but also the long-standing division between industrialized and developing countries—dwarf individual states, making national boundaries increasingly irrelevant. Moreover, "from the perspective of national markets there has been an effective de-concentration of capital" as cartels are dismantled and banks and industry separate (Lash and Urry 1987, 5).

The export of manufacturing activities to the Third World has transformed the occupational structure in the industrialized countries. The preponderance of financial activities and the growing tendency of the largest corporations to subcontract specialized functions have produced the huge service sector with its familiar socioeconomic repercussions: smaller, more responsive enterprises with flexible labor processes replacing Taylorist organizational structures, feminization of the work force, and fractionalization of the working class. The political ramifications of these shifts include the decline of class-based politics, party dealignment, and the delegitimation of state intervention, including the breakdown of corporatist and neocorporatist mechanisms of state regulation, wage negotiations, and national planning, as well as challenges to the welfare-state. Finally, disorganized capitalism is revealed culturally in the proliferation of postmodernism and spatially in the substitution of suburbs for cities as population and economic centers, and in the resurgence of localism. Although a global phenomenon, the authors suggest that national peculiarities continue to determine the rate and extent of the disorganization of individual nation-states, with the process most advanced in those countries that were least organized to begin with, beginning with the United States and Britain in the 1960s, France in the late 1960s, Germany in the 1970s, and Sweden in the late 1970s and early 1980s.

The key explanatory factor of disorganized capitalism is the "decline of working-class capacities" (Lash and Urry 1987, 11). Class capacities include not only the numerical size of the working class but also the organizational and cultural resources at its disposal. "Not only has the size of the working class and especially its 'core' declined in disorganized capitalism, but spatial scattering has meant the disruption of communicational and organizational networks, resulting in an important diminution of class resources" (Lash and Urry 1987, 11). Another important factor in the emergence of disorganized capitalism was America's organization of the global economy between 1945 and the 1970s based on the Bretton Woods Conference, which established the dollar as the global principal reserve currency, the extensive loans to Western Europe under the Marshall Plan, and American corporations' huge investments throughout the global economy. These actions eventually eroded the hegemony of Pax Americana as new centers of economic power emerged.

> In the case of the USA, its world-wide dominance spread capitalist relations throughout the world economy, particularly into the hundred or so countries which had previously been part of various imperial blocs, each linked into one or other of the organized capitalist societies. That in turn led in the 1970s to a number of disorganizing tendencies in the world economy which reproduced many of those features found in the USA since the late 1940s. (Lash and Urry 1987, 114)

The characteristics of disorganized capitalism can be summarized in three points: first, the internationalization of the world economy and the resultant effects on nation-states; second, the related changes in class structures within nations and most importantly, the rise of a service class that develops out of the relations of capital and labor in the organized era and becomes a powerful third player; and third, the cultural changes that result from the first two points, particularly the decline of working-class capacities, one result of which is the decrease in traditional class conflict–type social struggles and the increase in sectoral, special interest, and new social movement struggles. These three points now will be developed more fully.

American capitalism was thoroughly organized at the top in the late 1800s, when industries formed cartels and cooperated with banks regarding investments, and the U.S. state provided instrumental support. There was not such organization at the bottom during this period, and the "American polity in organized capitalism's first moment (from the 1890s) was characterized by the state apparently acting as the instrument of the economically dominant class. Subsequently, the "progressivism" of the New Deal helped American capitalism organize at the bottom and lent relative autonomy to the state" (Lash and Urry 1987, 10).

Progressivism is an ideology associated with the service class and related middle classes, and plays a crucial role in the disorganization of capitalism; progressivism "has always been the main source of opposition to unregulated capitalist accumulation in the USA" (Lash and Urry

1987, 10). Although the American working class never did fully organize at the bottom due to the successive waves of immigration and the resulting ethnically divided character of workers, its lack of organization was also due to the "early presence, size and access to organization of the American new middle classes, and especially the 'service class' " (Lash and Urry 1987, 10). "In the USA, as we have suggested above, partly because the service class possessed such extraordinary resources so early on to pose opposition to capital, the working class never had the type of organized capitalist structuring presence that it had in Europe" (Lash and Urry 1987, 13).

Although the class capacities of the working class had been decreased by spatial scattering and fragmentation, "the size and resources of the professional-managerial strata, or 'service class,' have enormously increased (Lash and Urry 1987, 11). The rise of the service class occurred as professionals created space for their own class formation through the expansion of universities and professional associations and through their successful arguments justifying their positions in terms of superior education and professional expertise—the emergence of Taylorist engineers contributed to this phenomenon. Lash and Urry's claim is that "the service class, which is an effect or outgrowth of organized capitalism, is subsequently, largely through its self-formation, an important and driving factor in capitalism's disorganization process" (1987, 11).

Class dealignment and the growth of "catch-all," non–class-based parties and related new social movements characterize postmodern culture and disorganized capitalism. It is the middle-class youth of the service class who make up the "new social movements" associated with postmodernism. Middle-class youth grow up with opportunities to avoid having their lives and identities structured by traditional categories of class. "Middle-class youth, then, and the expressive professions in the service class are a potential audience for postmodernist culture, and a potential source of resistance to domination in disorganized capital" (Lash and Urry 1987, 15).

The ascendance of these new social movements has been indicated by the breakdown of voting patterns along class lines.[17] Successes of the movements include tax reform, environmental protection, consumer protection, occupational health and safety, affirmative action, product liability, expansion of the welfare-state, and regulation of multinational corporations. During this era there was a marked increase in legislation adopted in opposition to corporate interests. In the United States between 1965 and 1975, "25 major pieces of legislation were passed regulating corporations in terms of consumer and environmental protection, occupational health and safety, and personnel policy" (Lash and Urry 1987, 222). According to Lash and Urry:

> Conservatism fails to get to grips with the emergence of new social groupings and forms of social organization which are not simply generated by or related to developments in the market-place. To put it slightly differently, the combination of the strong state and the unregulated market excludes and alienates many contemporary social forces and sets of social relations. (1987, 222–23)

Habermas (1981) distinguishes feminism from the diversity of other new social movements. Feminism, as opposed to the antinuclear, environmentalist, peace, or other movements, "seeks to conquer new territory" as opposed to defensive actions against the "formal 'organized' spheres of activity" (Lash and Urry 1987, 220). Feminism, therefore, is the only movement directly against patriarchal oppression.

The creation of national organized capitalism required the spatial concentration of the means of production, distribution, and social reproduction within national boundaries. The process of disorganization has meant a spatial scattering or deconcentration of socioeconomic relations based on national boundaries. This spatial scattering has resulted in the decline of the city, region, and nation-state and includes a process of

> first, the spatial deconcentration of the various production processes within today's large firm. Second, of the disurbanization of the means of production, not just to the suburbs and Third World subsidiaries but to the countryside in the First World. Third, the disurbanization of executive functions and of commercial capital. Fourth, the spatial scattering of the means of collective consumption, which has meant the residential deconcentration of labour power, of the working class itself. Finally, the growth of the highly capitalized establishment—in industry, commerce, and services—and the corresponding decline in number of employees per workplace has resulted in the spatial deconcentration of labour on the shopfloor. (Lash and Urry 1987, 10–11)

Although spatial scattering does imply a decentralization of production and a new spatial division of labor resulting in a tendency for smaller average plant size and an increase in the number of small enterprises, "this does not mean that large enterprises are not increasingly significant for the overall organization of western capitalism" (Lash and Urry 1987, 88). "In analyzing such transformations it is clear that a centrally important change has concerned the growth of very large corporations" (1987, 88).

Lash and Urry provide an ideal typical representation of the way that corporations develop in the three stages of capitalist society. During liberal capitalism all firms were small, there were high birth and death rates, and firms operated at the local or regional level. During organized capitalism three kinds of firms developed: small firms that were unlikely to grow larger, the regional national company that sometimes developed multinationally, and the multinational company that grew "monocentrically" so that the parent company kept tight control over its foreign subsidiaries and branches, a mother-daughter relationship. "In each case capital was nationally owned and clearly tied into the fortunes of the country in which it was owned and to which it was indissolubly bound" (1987, 89). In disorganized capitalism yet another kind of enterprise emerged. Some firms developed a "polycentric" structure and multidivisional organizational form as their subsidiaries began to operate more independently.

> Certain of these polycentric multinationals develop into the crucially important "global corporations" as they take over other firms which themselves are operating multinationally, firms often producing quite different products and based on different industrial processes. The attachment to any single economy becomes more tentative, as capital expands (and contracts) on a global basis. A much more complex spatial division of labor develops in which different parts of different production processes are separated off and develop within different national economies, depending, in part at least, on relative wage rates and worker organization. The spatial division of labour, which in earlier phases results from the unplanned patterns established by a considerable number of legally separated enterprises, becomes under disorganized capitalism a largely planned development internal to the vast global corporation. (Lash and Urry 1987, 90)

Large multinational corporations have emerged not only to facilitate the growth of mass-production technology, but also to orchestrate and regulate a given national economy (Lash and Urry 1987, 196; Piore and Sabel 1984). Disorganized capitalism's effects on these "global corporations" manifests as (1) a loss of control over mass markets for their products in the industrialized countries; (2) the ability to make up some losses by penetrating Third World economies; and (3) increased competition in all markets due to increased import penetration of each given national economy by other global corporations and decreased world economic concentration levels.

The British geographer David Harvey (1990) provides another detailed account of the transition from Fordism to post-Fordism. He uses the regulationist school's metanarrative to analyze the political-economic transformations of late twentieth-century capitalism. The development of Fordism after World War II created a "total way of life" characterized by workers, capital, and the state cooperating to maintain the stability of the system, a system where workers won considerable concessions and benefits. The global recession in the 1970s shattered the Fordist regime, which was replaced by a new regime based on flexible accumulation, flexible in its labor processes, labor markets, production arrangements, and consumption patterns. Harvey argues that flexible accumulation is just another aspect of capitalism that copes with the current accumulation crises by using strategies of temporal and spatial displacement to a greater degree than ever before. Although Fordism was characterized by rigid relationships based on national class compromises, flexible accumulation fractured the spatial fix of industrial cities or countries and replaced it with fragmented and unstable relationships. In particular, Harvey argues that the shift from Fordism to flexible accumulation has been characterized by a particular period of space-time compression associated with the electronic media and image management.

Harvey's interest is the complex web of dynamic interrelations that capitalist systems require to operate effectively over a period of time. These interrelations provide the institutional collective action necessary for sustained accumulation. Some degree of collective action is required to compensate for market failures, to prevent excessive economic concentration, and to provide collective goods. This collective action is usually provided by the state, an interventionist and regulatory state. Fordism did not come to fruition until the New Deal political apparatus was conceived and implemented to counter the near collapse of capitalism during the Great Depression and the World War II mobilization to defeat the nationalist socialist powers. The combination of the interventionist state and reliance on scientific methods in all aspects of production facilitated the institutionalization of large-scale planning and the rationalization of the work process. The capitalists feared the centralization of control in the state, and labor resisted assembly-line production, but "it was hard for either capitalists or workers to refuse rationalizations which improved efficiency at a time of all-out war" (Harvey 1990, 12). What was needed was a set of scientific managerial strategies complemented by state actions that would stabilize capitalism while at the same time avoiding negative repercussions of extreme nationalism. By the end of the war, the new modes and mechanisms of state intervention needed to match the requirements of Fordist production had been developed and deployed. Fordism had found its mode of regulation in Keynesianism and Bretton Woods.

The postwar boom was characterized by stable rates of economic growth and rising living standards in the United States. The phenomenal growth brought about by the postwar boom depended on a corporatist compromise by capital, labor, and the state. The state had to take on an interventionist role as the mediator of class conflict and as the institutional support for mass consumption. While social security, minimum wage, and unemployment insurance supported mass consumption, the Wagner Act of 1933 legitimized collective bargaining and the Taft-Hartley Act of 1952 purged the radical communists from the legitimized union movement. "The tense but nevertheless firm balance of power that prevailed between organized labour, large corporate capital, and the nation-state, and which formed the power basis for the postwar boom, was not arrived at by accident. It was the outcome of years of struggle" (Harvey 1990, 133).

Central to the corporatist class compromise was the purging of the radical union element and the agreement by labor to trade most of the control over the organization of work for wage gains. Intercapitalist competition was curbed through the centralization of capital and the emergence of oligopolistic pricing and planning, which facilitated the corporate-wide commitment to high fixed-cost investment in technological change and the development of large economies of scale through standardized mass production. Scientific management brought rationality and bureaucratization to corporate decision making. "The decisions of corporations became hegemonic in defining the paths of mass consumption growth, presuming, of course, that the other two partners in the grand coalition did whatever was necessary to keep effective demand at levels sufficient to absorb the steady growth of capitalist output" (Harvey 1990, 134).

To provide the stable demand conditions required for high fixed-cost investment in mass production to be consistently profitable, the state worked to moderate business cycles through programs of fiscal and monetary policy. Such efforts included public investment in transportation systems and utilities that were needed for the growth of a system of mass production linked to mass consumption, as well as welfare-related items such as social security, minimum wage, education, farm subsidies, and housing. The state managed this mix of various policies and programs that created the coordination required for the stable postwar accumulation regime. "Fordism depended, evidently, upon the nation state taking . . . a very special role within the overall system of social regulation" (Harvey 1990, 135).

During postwar Fordism, under the protection of U.S. economic and military hegemony, U.S. national capital also spread across the world as multinational corporations expanded into Europe, Asia, and the decolonized nations. The Marshall Plan and U.S. foreign direct investment implanted Fordism in many new parts of the world; this allowed U.S. production capacity surplus to be absorbed elsewhere in newly developed global, mass markets. The expansion of foreign trade and international investment "meant the globalization of the supply of often cheaper raw materials (particularly energy supplies)" (Harvey 1990, 137). Under Bretton Woods the U.S. dollar became the "world's reserve currency and tied the world's economic development firmly into U.S. fiscal and monetary policy. The United States acted as the world's banker in return for an opening up of the world's commodity and capital markets to the power of the large corporation" (Harvey 1990, 137).

Even at the apogee of Fordism, not everyone was included in the benefits of the postwar boom. Labor markets tended to be divided into the privileged "monopoly sector" of white males in unions and a marginalized "competitive sector" made up of a much higher proportion of women and racial and ethnic minorities (O'Connor 1973). The inequalities of income and benefits meant that those workers in the competitive sector did not enjoy the levels of mass consumption as did the monopoly sector workers. The civil rights and feminist movements were examples of the "strong counter-movements of discontent with the supposed benefits of Fordism" (Harvey 1990, 138).

It was the state's role to spread the benefits of Fordism to those discontented groups or suffer a loss of legitimation. This became increasingly difficult as the postwar boom began to decline. "The countercultural critiques and practices of the 1960s therefore paralleled movements of the excluded minorities and the critique of depersonalized bureaucratic rationality. All these threads of opposition began to fuse into a strong cultural-political movement at the very moment Fordism as an economic system appeared to be at its apogee" (Harvey 1990, 139).

Discontent was also growing in the Third World, where the promises of modernization brought oppression to the masses instead of emancipation. Those who did benefit from the incursion of U.S. MNC foreign direct investment were the few local elites who chose to cooperate with the yankee capitalists. Socialist and bourgeois nationalist movements rose up in response to the broken promises of global Fordism to threaten U.S. geopolitical hegemony. Even with the rising levels of discontent both nationally and internationally, the Fordist postwar boom held together until the recession of the early 1970s. Since then a "rapid, and as yet not well understood, transition in the regime of accumulation began" (Harvey 1990, 140). According to Harvey, by the late 1960s the recovered Japanese and European economies had saturated internal markets and began to pursue export markets in competition with U.S. firms, as well as to penetrate the U.S. market. Also, the efficiency of Fordist production methods had displaced more and more workers, resulting in a consumption decrease that was offset by the U.S. war on poverty and the Vietnam War. Declining corporate profitability combined with the "wars" created a fiscal crisis for the U.S. state, which responded with inflationary fiscal and monetary policies that undermined the role of the dollar as the stable international reserve currency. "The formation of the Eurodollar market, and the credit crunch of 1966–67, were indeed prescient signals of the United States' diminished power to regulate the international financial system" (Harvey 1990, 141).

It was also during the 1960s that peripheral Fordism and NICs emerged, as many developing countries (especially Latin American) began their import substitution programs, and multinational corporations expanded their operations into Asia and other areas. This new wave of Fordist industrialization based in regions where there was little or no social contract with labor intensified international competition to the point that U.S. hegemony was successfully challenged, the "Bretton Woods agreement cracked and the dollar was devalued. Floating and often highly volatile exchange rates thereafter replaced the fixed exchange rates of the postwar boom" (Harvey 1990, 141). For Harvey, this was the period when the Fordist regime of accumulation and the Keynesian mode of regulation lost their ability to contain the inherent contradictions of capitalism. The rigidity of (1) large-scale, long-term fixed capital associated with mass production that could not adjust easily to changing consumer markets; (2) labor markets, labor allocation, and labor contracts associated with the capital-labor accord and a strongly entrenched working class; and (3) welfare-state commitments to entitlement programs that were necessary to maintain legitimation, while corporate tax revenues were on the decline all contributed to the "dysfunctional embrace of such narrowly defined vested interests as to undermine rather than secure accumulation" (Harvey 1990, 142). The only tool the U.S. state had to combat the accumulation crisis and keep the economy stable was to print money, an action that increased inflation.

Loose monetary policies by the United States and Britain maintained the postwar boom until 1973. Efforts to stem the rapid increase in inflation in 1973 triggered a recession and a "worldwide crash in property values and severe difficulties for financial institutions" (Harvey 1990, 145). OPEC's decision to increase oil prices and the Arabs' decision to embargo oil exports to the West as a consequence of the West supporting Israel in the 1973 Arab-Israeli War increased energy costs to the industrialized world, and the resulting billions of petrodollars increased the instability of world financial markets. "The strong deflation of 1973–75 further indicated that state finances were overextended in relation to resources, creating a deep fiscal and legitimation crisis" (Harvey 1990, 145). Intensifying competition and high levels of excess capacity pushed corporations into a period of restructuring. Restructuring strategies included intensification of domestic labor control, geographical decentralization to regions of easier labor control, automation, mergers, technological change, and new niche market product lines.

The recession of 1973 combined with the "oil shocks" set into motion the current period of economic, political, and social restructuring.

> In the social space created by all this flux and uncertainty, a series of novel experiments in the realms of industrial organization as well as political and social life have begun to take shape. These experiments may represent the early stirrings of the passage to an entirely new regime of accumulation, coupled with a quite different system of political and social regulation. (Harvey 1990, 145)

For Harvey, flexible accumulation is the most interesting experiment.

Flexible accumulation "is marked by a direct confrontation with the rigidities of Fordism" and is characterized by flexible labor markets, labor processes, products, and patterns of consumption and by the emergence of new ways of financing services, new sectors of production, and "greatly intensified rates of commercial, technological, and organizational innovations" (Harvey 1990, 147). Flexible accumulation strategies have fostered shifts in the patterns of uneven development between and within geographic regions and the development of new industrial regions such as the Third Italy and Silicon Valley. Flexible accumulation has also brought about a new surge in "time-space compressions," as "the time horizons of both private and public decision-making have shrunk, while satellite communication and declining transport costs have made it increasingly possible to spread those decisions immediately over an ever wider and variegated space" (Harvey 1990, 147).

Faced with declining profits, increased competition, and market volatility, corporations used flexible accumulation strategies to increase their control over labor. Capital expanded production in regions characterized by weak labor organization and then, using the threat of capital flight, forced down wages and benefits domestically while initiating more flexible work regimes and labor contracts. There has been a move away from "regular employment" towards an increasing use of subcontracted, part-time, and temporary employment. Even regular workers are subject to "nine-day fortnights," when they work overtime during peak demand periods and shorter hours during slack times. Flexible accumulation implies "relatively high levels of 'structural' (as opposed to 'frictional') unemployment, rapid destruction and reconstruction of skills, and modest (if any) gains in the real wage, and the rollback of trade union power—one of the political pillars of the Fordist regime" (Harvey 1990, 147, 150).

The result is a labor market characterized by a shrinking core of full-time permanent employees who enjoy the benefits of high levels of consumption associated with core sector employment but who must be more amenable to flexibility and geographical mobility and an expanding periphery made up of two segments: full-time, but not permanent, employees with easily available standardized skills (such as clerical or secretarial) and less access to career ladders and higher turnover rates; and part-timers, casuals, fixed-term contract staff, temporaries, subcontractors, and public-subsidy trainees. The high turnover rates and standardized skills associated with periphery full-timers and the low levels of job security of the highly flexible periphery workers provide high degrees of numerical flexibility in matching employment levels with demand. "All evidence points to a very significant growth in this category of employees (periphery) in the last few years" (Harvey 1990, 151).

Transformations in industrial organization have accompanied labor market shifts. Capital's increased reliance on subcontracting has facilitated the reemergence of small businesses and of "sweatshop" forms of production often based on "older systems of domestic, artisanal, familial (patriarchal), and paternalistic ('god-father,' 'guv-nor,' or even mafia-like) labour systems" (Harvey 1990, 152). There has also been a rapid growth in the informal, black, and underground economies throughout capitalist countries. All of these systems are not amenable to labor organization or collective class struggle. "Struggling against capitalist exploitation in the factory is very different from struggling against a father or uncle who organizes family labour into a highly disciplined and competitive sweatshop that works to order for multinational capital" (Harvey 1990, 153).

The effects of these new systems of industrial and labor organization are especially detrimental to women. Flexible labor market structures often substitute lower paid female workers for the highly paid core male workers as well as subject them to a resurgence of patriarchical practices under subcontracting and homeworking. The multinational strategy to take Fordist mass-production systems to Third World production sites and exploit women's labor under conditions of low pay and low job security is exemplified by the Maquiladora system on the U.S.-Mexico border (e.g., Nash and Fernandez-Kelly 1983). The Maquiladora system allows U.S. capital and managers to remain north of the border while locating factories that employ mostly young Mexican women south of the border.

The effects of these new systems of industrial and labor organization have also negatively impacted traditionally organized businesses, resulting in a wave of restructuring, deindustrialization, bankruptcies, and plant closures. Mass-production systems have tended to migrate to low-cost production sites in the Third World, while firms and industries that were able to converted to flexible production, "with its emphasis upon problem solving, rapid and often highly specialized responses, and adaptability of skills to special purposes" (Harvey 1990, 155). The composition of the top *Fortune* 500 corporations has changed dramatically during the past twenty years, including a stagnation in their percentage of global employment. During the same time period, new business formation increased rapidly, as many small businesses "inserted themselves into the matrix of sub-contracting skilled tasks or consultancy" (Harvey 1990, 155).

Modern industrial organization systems are made up of a mix of Fordist mass-production systems and flexible production systems with the capacity to manufacture a variety of goods in small batches. Often the new organization of industry is characterized by

> the rise of entirely new industrial forms or . . . the integration of Fordism with a whole network of sub-contracting and "outsourcing" to give greater flexibility in the face of heightened competition and greater risk. Small-batch production and sub-contracting certainly had the virtues of bypassing the rigidities of the Fordist system and satisfying a far greater range of market needs, including quick-changing ones. (Harvey 1990, 155–56)

Harvey disagrees with the arguments of Offe (1985) and Lash and Urry (1987) regarding the increasing disorganization of capital. Instead of the trend toward increasing disorganization, Harvey argues that the traditional tensions within capitalism between monopoly and competition and between centralization and decentralization are being worked out in new ways that result in capitalism "becoming ever more tightly organized through dispersal, geographical mobility, flexible responses in labour markets, labour processes, and consumer markets, all accompanied by hefty doses of institutional, product, and technological innovation" (Harvey 1990, 159). This "more tightly organized capitalism" has been achieved by the development of information as a highly valued commodity. Instant data analysis of accurate and up-to-date information provides crucial market signals for flexible small-batch production responding to changes in fashion or tastes. Privileged access to information also enables instantaneous corporate responses to changes in exchange rates, business moves by competitors, scientific discoveries, political upheavals, or government policies. "Knowledge itself becomes a key commodity, to be produced and sold to the highest bidder, under conditions that are themselves increasingly organized on a competitive basis" (Harvey 1990, 160).

As a result of the increased importance of information, organized knowledge production on a commercial basis has expanded significantly in the past few decades. There has been a transition in many university systems from "guardianship of knowledge and wisdom to ancillary production of knowledge for corporate capital" (Harvey 1990, 160). This trend has been accompanied by a concentration of control over information flow and over the "vehicles of popular taste and culture," such as book publishing and advertising, which "have likewise become vital weapons in competitive struggle" (Harvey 1990, 160).

More importantly, there has been a "complete reorganization of the global financial system and the emergence of greatly enhanced powers of financial coordination" (Harvey 1990, 160). This has been a dual movement characterized by an emergence of powerful global financial conglomerates as well as a formation and proliferation of decentralized financial activities through entirely new financial arrangements.

> Deregulation and financial innovation—both long and complicated processes—had by then become a condition of survival of any world financial center within a highly integrated global system coordinated through instantaneous telecommunications. The formation of a global stock market, of global commodity (even debt) futures markets, of currency and interest rate swaps, together with an accelerated geographical mobility of funds, meant, for the first time, the formation of a single world market for money and credit supply. (Harvey 1990, 161)

The new structure of the global financial system has produced a blurring of boundaries between the previously distinctive services of banking, brokerage, financial services, housing finance, consumer credit, and commodity, stock, debt, and currency futures. Industrial capital is now "so integrated into financial operations and structures that it becomes increasingly difficult to tell where commercial and industrial interests begin and strictly financial interests end" (Harvey 1990, 161). Part of the confusion is associated with the growth of "paper entrepreneurialism" and its focus on making money in other sectors besides services and the production of goods. "Paper entrepreneurialism" includes creative accounting, corporate raiding and asset stripping, and profiting from shifts in currency exchange rates and interest rates.

Harvey presents the case of the growth in "stateless" Eurodollar markets that are "quite uncontrolled by any national government" as an example of how the world's financial system has "eluded any collective control on the part of even the most powerful advanced capitalist states" (1990, 163). Whereas under the Fordist mass-production system industrial capital provided the coordinating system of funding, flexible specialization relies on a closer relationship with finance capital. Finance capital is more fluid and flexible in time and space as opposed to industrial capital with high fixed costs in production and consumption. "The increasing powers of co-ordination lodged within the world's financial system have emerged to some degree at the expense of the power of the nation-state to control capital flow and, hence, its own fiscal and monetary policy" (1990, 164).

THE AMERICAN RADICALS

The American radicals can be divided up into two groups: those who concentrate on social structures of accumulation (e.g., Gordon et al. 1982), and those who concentrate on technological innovations and the associated organization of work paradigms (e.g., Piore and Sabel 1984).

The most distinctive group is those who use the concept of "social structures of accumulation" (SSA), first advanced by Gordon (1978, 1980). This approach argues that the current crisis is linked to the decay phase of the SSA associated with the current long wave of capitalist development. Sustained periods of accumulation require specific social and political arrangements conceived as SSAs to support and reinforce the economic factors that facilitate growth. SSAs include the political, legal, labor relations, and marketing arrangements under which individual companies operate; Marxists call this the superstructure. Different SSAs are produced in and through a specific balance of forces, and changes in this balance can cause major economic ruptures. In essence, SSAs are similar to the concept of mode of regulation but put more emphasis on changes in the balance of power than do the European regulationists (Jessop 1990).

Gordon first argued that "relative stability in the general social and economic environment affecting the possibilities of capital accumulation is a necessary condition for sustained and rapid accumulation; without which such structural stability, the pace of capital accumulation in capitalist society is likely to slacken" (1980, 12). Gordon then presented a list of institutional complexes that must be present for capital to be able to overcome the intrinsic contradictions of class struggle and competition in capitalism and to accumulate effectively. These institutional complexes ranged from systems of natural resource supply and the "social family structure" of schools and family through labor markets and structures of labor management to structured social foundations of consumer demand (Gordon 1980, 12–17). The institutional complex formed a unified and decomposable SSA with its own logic and internal contradictions (Gordon 1980, 18).

For Gordon, it is the lack of an appropriate SSA to organize and stabilize capitalist accumulation that has fostered the current global economic crisis. Although cyclical business downturns might be restored by cyclical upswings, economic crises are periods of "economic instability in capitalist economies whose resolution depends on the reconstruction of a social structure of accumulation" (Gordon 1980, 20; see also Gordon et al. 1982, 26). The reconstruction of a new SSA only occurs after a long class struggle, with the shape of the SSA reflecting the differing strengths of the competing interests (Gordon 1980, 21–22).

Much of the SSA work has been focused on the postwar boom in the United States and the characteristics of the SSA that facilitated that boom. Four broad institutional complexes characterized that era: (1) the capital-labor accord that brought big capital and organized labor together in a class compromise in which labor gained the right to unionize and rising wages tied to rising productivity and capital gained extensive control over the organization of work; (2) the international balance of forces was maintained by Pax Americana and the U.S. dominance in the supply of raw materials and world markets; (3) the capital-citizen accord; and (4) competition between domestic capitals (Bowles and Gintis 1982; Bowles et al. 1988; Gordon et al. 1982). "Together these accords produced a balance of power favorable to accumulation and subsequent shifts explain the movements of profit rate within the 'postwar corporate system' " (Jessop 1990, 192).

Gordon and his associates (1982) reinterpret the history of U.S. industry and labor within a periodization of labor market divisions and their associated managerial control systems. The historical weakness of the labor movement in the United States and the resulting failure of socialism to successfully develop are results of capital's strategies to keep the working class fragmented. According to Gordon et al., "The disunity of the U.S. working class persists in large part as a result of the objective divisions among workers in their production experiences; these objective divisions constitute both a consequence of continuing capitalist development in the United States and a barrier to a unified anticapitalist working-class movement" (1982, 8).

Workers are divided, not only by their daily circumstances, but by the whole pattern of their working lives: opportunities for training and advancement, systems of managerial control, job security, working conditions, and renumeration, all varying greatly from sector to sector. Workers' interests and the strategies they use to secure them often diverge from one sector to another. Furthermore, labor market segmentation and racial and sexual discrimination reinforce one another so that inequality is deep-rooted and routinely perpetuated. These factors, combined with ethnic and cultural divisions, produce a fragmented working class unable to coalesce politically or to successfully develop its own programs.

Strands of Marxian economic analysis form the foundation of these authors' conceptual framework. They use the "long waves" theory of Kondratieff cycles to explain the regular emergence of the internal contradictions of capitalist development. They link the ebbs and flows in the power of American labor to the long-wave cycles in economic expansion and contraction and to a series of crises in the history of American capitalism.

U.S. history is characterized by three stages of capitalist development and their associate crises: proletarianization, homogenization, and segmentation. Each of these stages has ended when its SSA could no longer provide the required stability for accumulation. After a long period of experimentation with methods of industrial organization, new arrangements between capital, labor, and the state emerged and became the prevailing SSA.

Proletarianization, from the 1820s through the 1890s, was the period when the work force of wage laborers was created. Capitalist entrepreneurs financed and organized production and marketed their products but the shop floor was mostly organized by the skilled workers based on craft lines. By the late nineteenth century, however, employers increasingly sought to break the power and appropriate the knowledge of these same skilled workers in order to introduce new production techniques, weaken craft unions, and lower production costs.

The period of homogenization, from the 1870s to World War II, saw Taylorist management practices combined with mass-production techniques to reduce skill levels required from labor. Anti-union campaigns, the subdivision of tasks, and machine-regulated or assembly-line mass production were all used by capital and management to create a mode of production that for the most part required only semiskilled workers. The homogenized semiskilled workers responded to capital's attack on craft unionism by organizing into industrial unions; the worker had been deskilled but the unions had adapted and survived. Capital needed another strategy to destabilize the unions.

The current period of labor segmentation, from the 1920s to the present, is characterized by capital's strategy to divide the work force while keeping core-sector workers in line by offering incentives (profit sharing) and promotion ladders (often worked out in collective bargaining) and by introducing new technologies that set the pace and measure the output of their operators. These core industry workers are employed in large, capital-intensive, monopolistic firms that employ long-tenured, well-paid, unionized workers—in the United States almost always white males. The rest of the workers on the periphery are employed in small, highly competitive firms that operate in unstable markets and employ transient, low-paid, nonunion workers—often women and minority groups.

In each era capitalism reaches the limits of its existing structure; as labor productivity declines, employers are unable to discipline their workers, and crisis permeates the economy. The internal contradictions of capitalism cause employers to search for new methods of labor discipline during the period of crisis (the contraction wave in the Kondratieff cycle), which reaches its fruition during the next wave of economic expansion.

Gordon et al. (1982) maintain that the U.S. economy entered into the current period of crisis in the beginning of the 1970s; the symptoms of this crisis were the OPEC oil shocks, declining productivity, deficit spending by the welfare-state, and an unregulated international monetary system. The postwar prosperity grew out of the "integrated complex of law, institutions, and customary arrangements that we have labeled the 'postwar accord' " (1982, 240). The SSA associated with the postwar accord was

> rooted in the Wagner Act, Social Security, Taft-Hartley, the employment Act of 1946, and the Bretton Woods system and encompasses segmented labor markets, diverse (simple, technical, and bureaucratic) systems of control of the labor process, class conflict channeled into the governmental arena, an extensive economic role for government, and the maintenance of a hegemonic military posture to protect the opportunity for American corporations to invest abroad. These institutional arrangements have begun to decay. (1982, 240)

By the end of the 1960s, the postwar accord and associated SSA began to unravel and thereafter could not maintain the necessary levels of stability for accumulation. The results of channeling class conflict into the state arena produced a dramatic increase in the role and costs of the state as it tried to mediate class confrontation, resulting in an accumulation and legitimation crisis for the state. The failure of the United States to police the capitalist world in Vietnam prompted the rise of other Third World anticapitalist insurgencies; the best example being OPEC. The growth of powerful economic capitalist competitors in Europe and East Asia "eroded the dominant and privileged economic position of the United States, undermining the basis of Bretton Woods arrangements" (Gordon et al. 1982, 241). As economic growth slowed in the United States, firms decentralized production; first, from the northeastern-midwestern rustbelt to the southern sunbelt, then on to NICs and Third World countries, "where the term of the postwar accord did not obtain" (Gordon et al. 1982, 241). The resolution of the crisis would necessitate a new SSA, requiring the creation of a new political constituency to effect the necessary changes.

> Construction of a new social structure of accumulation requires major institutional innovations, and such innovations require for their implementation a new political agenda and new sources of

> support—in other words, the creation of a new political coalition. The onset of a long-swing boom is accompanied by the emergence of a new dominant political force, and their joint appearance is significant. The new political group (electoral party or wing of an established party) proposes the needed institutional innovations and, if it is successful in implementing them and if the reforms are in turn successful in stimulating the boom, the new political force reaps the credit. Its ideology becomes the "dominant" ideology ("the business of America is business" or "mixed capitalism" in the welfare/warfare state), and the political faction becomes the dominant political force. (Gordon et al. 1982, 242)

The authors present the rise of the New Right under the Reagan administration as a possible precursor to the new SSA. The Reagan program attempted to "revive the boom by rolling back the gains achieved by working-class groups under the postwar accord, as though the years of postwar prosperity could be run through again once environmental laws, health and safety regulations, income maintenance programs, and the rest of the welfare state apparatus have been dismantled" (Gordon et al. 1982, 243). Whether the Reagan agenda succeeds will be determined by the ability of the "opposition" to advance a coherent alternative plan. For these authors, their hope lies with the "Left" and labor who must advance a successful plant for the construction of a new SSA lest the rise of the New Right's SSA ensures continued "widespread hardship and risks social catastrophe. This kind of opportunity for restructuring comes only once in a generation" (Gordon et al. 1982, 243).

Gordon (1988) provides a critique of the proponents of the globalization of production (GOP) and new international division of labor (NIDL). He argues against fatalistic conclusions that assert that these twin tendencies "manifest such deep structural transformations in the world economy that group or government efforts to swim against the currents are becoming increasingly ineffectual, if not futile. The power of labour, community and the state has seemed to wither as multinational corporations sweep irresistibly around the globe" (1988, 24).

Gordon admits that rising unemployment, sectoral devastation in many traditional industries, and corporate demands for concessions on wages, benefits, and working conditions clearly warrant concern. But he attributes these effects to the continuing stagnation of the world capitalist economy and not to a transformation of the global capitalist economy. According to Gordon: "These changes are best understood 'not' as a symptom of structural transformation but rather as a consequence of the 'erosion' of the social structure of accumulation which conditioned international capitalist prosperity during the 1950s and 1960s. We are still experiencing the decay of the older order and not yet the inauguration of the new" (1988, 25). Gordon writes "from the urgent concern that we not handcuff ourselves politically with misplaced and potentially crippling perceptions about a new global order" (1988, 26). He further states that, although he shares the general concern over the enhanced power of capitalists and their "corporate juggernauts," they are "staggering from the blows" of the current economic crisis. Although TNCs have "huge resources" and "they can still take their capital out on strike if they dislike our behavior at the bargaining table," he maintains that it is "not at all clear that they have already achieved new global structures of coordination and control which will 'necessarily' enhance the power through the coming decades" (1988, 25). TNCs are much more vulnerable and much more uncertain about the future than GOP and NIDL proponents admit.

Gordon systematically reinterprets the dominant tendencies of the GOP and NIDL thesis. First, he argues that we have not witnessed an increasingly "open" international economy with "productive capital buzzing around the globe" but rather have moved rapidly toward an increasingly "closed" economy for productive investment. Production and investment decisions are increasingly dependent upon a range of institutional policies and a pattern of differentiation and specialization among the less-developed countries (LDCs). Whereas the LDCs were once wide open for foreign direct investment and resulting exploitation, access to abundant raw materials and low-cost labor currently is complicated by nationalist strategies and political instability. "The international economy, by the standards of traditional neoclassical and Marxian models of competition, has witnessed 'declining' rather than 'increasing' mobility of productive capital" (1988, 63).

Second, contrary to the assertion that state policies have become increasingly futile on the international front, Gordon argues that the role of the state has grown substantially since the 1970s. The state has had to become more involved in active management of monetary policy and interest rates in order to mediate exchange rate fluctuations and short-term capital flows. In the era of monetarist conservatism, "everyone including the transnational corporations has become increasingly dependent upon coordinated state intervention for restructuring and resolution of the underlying dynamics of crisis" (1988, 64).

Gordon thinks that his clarifications of the changing global economic order help avoid the spreading political fatalism in the advanced countries, a fatalism that believes that "if we struggle to extend the frontiers of subsistence and security at home, one gathers, we shall stare balefully at capital's behind, strutting across the continents and seas, leaving us to amuse ourselves with our unrealized dreams of progress and the reality of our diminishing comparative advantage" (1988, 64). Although the postwar system of socially determined, stable institutional relationships is rapidly eroding, TNC responses of the 1970s

> reflect their own political and institutional efforts to erect some shelters against the winds of spreading economic instability. The TNCs are neither all-powerful nor fully equipped to shape a new world economy by themselves. They require workers and they require consumers. Workers and consumers helped shape the structure of the postwar system, and we are once again in a position to bargain over institutional transformation. The global economy is up for grabs, not locked into some new and immutable order. The opportunity for enhanced popular power remains ripe. (Gordon 1988, 64)

The second grouping of American radicals is found in the crossnational works of Piore and Sabel (Piore and Sabel 1984; Sabel 1982) and others who further elaborated their original ideas (e.g., Hirsch and Zeitlan 1988, 1991). Adopting a normative posture, for these authors flexible specialization should be the next regime of accumulation. Flexible specialization is advanced as a new form of production consisting of an emerging relationship between new technologies, new patterns of demand, and new forms of social organization of production that could become the "foundations of the new social democratic project in reconciling the interests of capital in securing high rates of productivity with the interests of labor in combining fulfillment at work and rising levels of incomes" (Clarke 1990, 75).

Piore and Sabel argue that we are at a historical moment where the new trajectory of capitalist work relations and technology is being established. They argue for the resurrection of the craft system of production in the form of flexible specialization, which combines the new computer technology of numerically controlled machines with the new consumer demand for non-mass-produced products and a supportive neocorporatist state. Numerically controlled machines require more skilled workers than mass-produced machines, therefore restoring some of the mental work and attachment to the job lost in Fordist production. The new technology provides the high rates of accumulation and less resistance from workers needed for a stable regime of accumulation. The authors cite examples of successes of flexible specialization systems pioneered by Toyota from Japan, the high technology sector in Baden-Württemberg, West Germany, and the Third Italy.

Sabel's original work (1982) on Italy related the new forms of high technology and cooperative artisan production to the historically specific economic, social, and political context in which they blossomed. There emerged in Italy in the 1970s a flexible network of regional districts of small and medium-sized firms that increasingly were using computer-assisted numerically controlled production to adapt to rapidly shifting markets. These new shops originally emerged as a product of the radical decentralization of production in the face of strike waves in the 1960s. The subcontractor shops began to federate, and they devised innovative products and processes that gave them increasingly independent access to global markets. Steel mini-mills, the farm and construction equipment industry, machine toolmakers, and ceramic makers established geographically specific arrangements supported by local government and Italian familial culture. Local government often subsidized industrial parks and research centers and provided for worker safety protection. Many of the new firms were family operations.

Piore and Sabel (1984) argue that the current global economic crisis is the "Second Industrial Divide," where the technology of computer-assisted numerically controlled machines could provide a new era of prosperity. They assert that there are two competing and complementary paradigms for organizing and managing production: mass production and craft production. Mass production replacing craft production was the first industrial divide, but mass production cannot provide accumulation when markets are specialized and shifting. They assert that within the proper institutional environment, craft production based on flexible specialization facilitates high levels of accumulation by adapting to changing consumer tastes and avoiding labor unrest.

Starting in the late 1960s, a series of five developments combined to plunge the mass-production economy into crisis: (1) social and labor unrest, (2) raw materials shortages, (3) rapid inflation, (4) rising unemployment, and (5) economic stagnation. These five developments, combined with the saturation of national mass markets, the worldwide drive for international sales and production, the rise of specialty markets, and the ineffectiveness of Keynesian state responses that cannot regulate a world where commodity and money flows refuse to recognize national boundaries, signaled the need for an alternative production system. Piore and Sabel argue that Silicon Valley, Route 128 surrounding Boston, the American steel mini-mills, and numerous similar examples from Japan and Europe highlight the success of companies that found profits in small, specialized, fast-changing markets that the mass producers could not service. They also argue that the attempts of mass-production firms to resolve the crisis through decentralization and subcontracting in the Third World, such as the "world car" program and "sweating" their workers, have not produced the efficiencies predicted by adherents to the mass-production paradigm.

The corporate response to the accumulation crises of mass-productionist economic organization first took the form of conglomeration, then multinationalization. Conglomeration was characterized by corporations trying to hedge their risks in their primary market by diversifying into other markets. This strategy did not work because the "risks it [corporations] sought to contain could not be reduced through diversification. For the risks arose not from business accidents, randomly distributed across markets, but from shocks to the economy as a whole. The result was a strong correlation among the problems of individual markets. Thus, risks in different markets were cumulative, not offsetting" (Piore and Sabel 1984, 197).

Realizing that the risk to be avoided was generated by macroeconomic shocks that disrupted the entire global economy, corporations turned to the strategy of multinationalization. Although macroeconomic strategies to resolve the accumulation crisis centered on strengthening the national institutions of neo-Keynesian regulation, multinationalization was a microeconomic strategy that sought to achieve economies of scale on a global basis that were unavailable at the national level through the extension of production to "many national markets simultaneously" (Piore and Sabel 1984, 198). The American automobile firms' strategy of the "world car," an extension to the global scale of Sloan's GM strategy, typified the strategy of multinationalization. Corporations attempted to maintain their domestic market positions while at the same time capturing portions of markets in developing countries to enable them to expand their economies of scale based on mass production. This attempt generally failed because there was no international regulatory regime to provide the stability necessary for coordinated transnational mass production and mass consumption.

Piore and Sabel (1984, 200–202) advance four hidden costs as the reasons the multinationalization strategy did not work. First, although dispersion of production to developing countries did originally avoid both the high costs of labor and labor unrest characteristic of developed economies, many of the new assembly-line workers quickly became discontented and sought improved workplace arrangements. Also, developing states that promised easy access and state support for industrial location of multinational subsidiaries often also made costly demands of the subsidiaries' revenues once they were established. To counter these tendencies, the corporations built identical plants in different countries to facilitate "multiple sourcing" and "playing one off against another," but this strategy was expensive "because the more the corporation multiplied its sources, the less it was able to take advantage of the economies of scale of global production" (Piore and Sabel 1984, 200–201).

Second, the geographically dispersed systems of parts provision and assembly proved inefficient in comparison to the Japanese "kanban" system of just-in-time production. Whereas the Japanese system required minimum inventories, the U.S. world car approach required large, costly inventories to hedge against interruptions of supply. Third, although the world car strategy was based on the standardized transnational production of a small car in response to the oil shocks of the 1970s, demand in the United States fluctuated between small and large cars depending on the price of oil and gasoline. This fluctuation made it much more difficult for world car strategists to consolidate a global market for small cars without the help of macroregulatory regime. Fourth, fluctuations in exchange rates and general international instability such as "unpredictable labor unrest, state regulation, and interruptions of supply" made the impact of foreign competition and the level of world demand uncertain (Piore and Sabel 1984, 201–2).

According to Piore and Sabel (1984), the strategies various national corporations used to counteract the crisis were grounded in the way the mass-production organization had initially emerged in the national context. The manufacturing sector in each country had adopted "the mass-production model in a nationally specific form, shaped by the manufacturer's relation to the world market, to the state, and to the postwar labor movement" (Piore and Sabel 1984, 222). In some countries the craft production system survived, while in others it almost disappeared. As the 1970s brought on increased economic uncertainty, the degree to which the residual craft paradigm persisted in each country played a large role in shaping corporate responses to the crisis.

> For the presence, if unacknowledged, of these craft elements served to obstruct potential solutions based on elaboration of the mass-production model; it foreshadowed—in a limited, intuitively if not yet theoretically comprehensible way—the possibilities of an alternative form of organization. Indeed, the discovery procedures that firms and national economies applied to the search for a solution were largely the institutionalized result of historical developments. Thus, any given country's residual barriers to wholesale Americanization served to guide local efforts at economic survival. (Piore and Sabel 1984, 222)

In places like Italy, Germany, and Japan, historical experience led the way to the successful reemergence of the craft paradigm and flexible specialization, resulting in increased accumulation; in the United States and France where there was little residual craft experience, attempts at economic reorganization stalled.

Piore and Sabel see two possible paths emerging from the current crisis. One route requires the construction of a new world economic order called "international Keynesianism," where mass production continues to dominate in the context of an amalgamation of markets divided into regional trade blocks, coordinated fiscal and monetary programs, a strengthened role for the International Monetary Fund and General Agreement on Trades and Tariffs that would provide Bretton Woods–like stability, and a new set of understandings between the developed and less developed nations that would balance global supply and demand and apportion economic growth around the world (1984, 252–57).

The other route is represented by flexible specialization or "yeoman democracy," where the institutions of management, labor, and government are reinvented to support the infusion of technology into the workplace so as to promote more flexibility in production and greater skills in the work force. Under this route, new social relationships would arise, reducing the distance between worker and boss and encouraging the introduction of technology that enhances the creativity of the workers. These new relationships would be regulated by a neo-Keynesian state similar to the New Deal (1984, 258–76).

Piore and Sabel conclude that it is possible that "flexible specialization and mass production could be combined in a unified international economy . . . [where] the old mass-production industries might migrate to the underdeveloped world, leaving behind in the industrial world the high-tech industries" (1984, 279). Although such a hybrid system would not last forever, "it would create a universal interest in two basic goals: worldwide prosperity and a transnational welfare state" (1984, 280).

3

GLOBAL POST-FORDISM AND TRANSFORMATIONS OF THE STATE

The literature introduced in Chapter 2 discusses the central role played by the state in Fordism. It was the intervention of the Fordist state that guaranteed the maintenance of key equilibria in the system and allowed both MNCs and their political representatives and subordinate (i.e., nondominant) groups to foster their interests. Indeed, in Fordism, the domestic state maintained the conditions for capital accumulation and simultaneously legitimized this action by allowing subordinate groups (such as workers) to increase their well-being in society. In this respect, it is widely agreed that the expansion of the welfare-state has been a key characteristic of the Fordist period. In general, the state has played a key role in the evolution of the agricultural and food sector (Friedland 1994; McMichael 1994). In global post-Fordism, it has been a key player in the creation and restructuring of food and fishery chains by fostering the interests of TNCs while simultaneously undergoing a process in which its powers and scope of action have been redefined.

For instance, one of the many examples of this trend is the work of Skladany and Harris (1995) on the global shrimp industry. Building on the research of Bailey (e.g., 1988) regarding the "Blue Revolution" and of McMichael pertaining to the transformation of the state in Global post-Fordism (e.g., 1994), they illustrate how fragmented elements of the state fostered the interests of both TNCs and nationally based companies in the exploitation of coastal zones for pond shrimp production. More specifically, through the actions of the state, TNCs and other relevant corporate players were able to redefine aquatic natural resource property relations away from traditional multiusages carried out by indigenous people. These restructurings primarily involved the creation of global circuits characterized by the insertion of intensive monoculture shrimp production into commodity chains that on the one hand provided cheap exports to affluent markets and on the other created significant environmental and social damages to developing regions of Latin America and Southeast Asia.

As a result of these restructuring processes, Skladany and Harris (1995) point out the changing character of the role of the state in global post-Fordism. They argue that because TNCs and national firms predominantly direct the shapes and trajectories of this new global regime, state power is significantly redefined. "Increasingly, the role of the state can be viewed as a mere backdrop against which rapid capital flows, international finance, technoscientific sophistry, environmental degradation, and social injustice is perpetuated by highly mobile TNCs and NCs [national companies]" (1995, 185). More specifically, because these firms are highly mobile and utilize a "slash and burn" pattern of pond shrimp development, "the ability of the nation-state effectively to 'manage' these contradictions is simply overwhelmed by the international scope of the industry's fluid structure which respects neither region or context" (1995, 22).

Discussing the issue of the state in the emerging global system, Friedland (1994) defines it in terms of four interrelated roles: maintenance of processes of accumulation of capital, maintenance of processes of social legitimation, mediation among various conflicting societal groups, and social and physical reproduction. He concludes that Fordism could not be reproduced without the steady presence of an interventionist state. It is also recognized by Friedland and others that the state played contradictory roles. Contrasting earlier and orthodox Marxist accounts, students of the Fordist state emphasize its repressive as well as emancipatory roles. In this respect, the Fordist state maintained the repressive characteristics of a "class" state as it reproduced the modes and forms of accumulation typical of capitalism (Offe 1985; Poulantzas 1978). Simultaneously, it allowed the introduction of legislation in favor of subordinate classes that protected interests contrary to those of dominant groups. As indicated in Chapter 2, the social and economic costs associated with a pro–subordinate classes interventionist state were one of the reasons for the collapse of the Fordist regime and the search for alternative and flexible organizational forms. Moreover, "accounts" of the future evolutions of the global post-Fordist regime pay particular attention to the characteristics of the post-Fordist state and to the manners in which the Fordist roles of the state will be transformed in post-Fordism.

The wealth of arguments on the centrality of the state in Fordism and its renewed importance in global post-Fordism mandate a review of pertinent literature. The objective of this chapter is to review literature on the state in order to facilitate the analysis of the tuna-dolphin controversy. The chapter reviews (1) traditional theories of the state, (2) transnational theories of the state, and (3) regulationist theories of the state.

TRADITIONAL THEORIES OF THE STATE

The role of the state in society has recently become a central topic in sociological debates. Generally, these debates cast the state as (1) an institution that is the direct instrument of the capitalist class (Miliband 1969, 1970; Domhoff 1967, 1979; O'Connor 1973, 1986), (2) an institution endowed with relative autonomy from capitalist domination (Block 1977, 1980; Offe and Ronge 1979; Poulantzas 1978), (3) an institution possessing almost total autonomy (Levine 1987, 1988; Skocpol 1979, 1985), and (4) an institution exhibiting a historical mix of instrumentalist and autonomous roles (Campbell and Lindberg 1990; Friedland 1983, 1991b; Hooks 1990; Jenkins and Brents 1989; Prechel 1990). The instrumentalist and relative autonomy perspectives come from the Marxist tradition, while the state autonomy perspective is from the Weberian tradition.

The Instrumentalist View

Bonanno (1991) provides a useful review of instrumentalist theories. The instrumentalist view argues that the state in capitalism is either "an instrument for promoting the common interests of the ruling [capitalist] class" (Offe and Ronge 1979, 346) or "a committee of the ruling class directly manipulated by the members of this class" (Carnoy 1984, 214). Instrumentalists take seriously Marx's contention that the modern state is but the "exclusive political sway" of the ruling class (Marx and Engels 1955, 11) and have sought to demonstrate how the capitalist class politically intervenes to assure its own interests. The instrumentalists emphasize the direct control that the ruling class exerts in all fundamental aspects of society, including the economy and polity.

There are generally two types of instrumentalist theory. The camp of Miliband (1969, 1970) and Domhoff (1967, 1979) argues that the state bureaucrats are part of the ruling class and are effectively bound to the ruling class through common socialization and educational backgrounds. This view contends that it is possible that members of other classes may enter the ruling capitalist class, but it is the latter that controls the state apparatus. Another instrumentalist view is advanced by O'Connor (1973, 1986). "State monopoly capital theory" asserts that monopolistic-corporate fractions of the capitalist class exercise direct control over the state by means of the control that they exercise over the economy (Bonanno 1991).

The instrumentalist approach has been criticized by studies from the relative autonomy approach that have emphasized the complex character of the relationship between the economy and the polity. Discrepancies between the actions of the state and that of the capitalist class cast doubts on the ability of the latter to directly control the former.

The Relative Autonomy View

The relative autonomy approach to the state asserts the partial independence of the state (superstructure) from the economy (structure) (Block 1977, 1980; Offe and Ronge 1979; Poulantzas 1978). The authors draw from the works of Marx (1947), Gramsci (1971, 1975), Habermas (1985), Horkheimer (1974), Lukács (1971), and Marcuse (1964). This approach emphasizes the role of ideology and the state in the development of capitalism. The state reproduces class relations not because it is directly controlled by the capitalist class or fractions of that class, but because the state is interested in reproducing the social relationships that support the dominant socioeconomic system (Offe and Ronge 1979).

Proponents of this approach (Block 1977, 1980; Offe 1985; Poulantzas 1978) suggest that the independence of the state from the capitalist class derives from its ability to mediate the short-term interests of class fractions and, simultaneously, to ensure the continuation of capitalism as a dominant mode of production. At the same time, however, the state can only favor the interests of capital to a certain degree because of its need to legitimize its actions. Legitimization involves maintaining a consensus of political strategies that are conducive to the maintenance of the status quo.

Offe (1985) contends that the inability of capitalism to maintain adequate levels of economic growth generates crises of accumulation that increasingly require the intervention of the state. Such interventions tend to penetrate new and expanding spheres of action outside the state's traditional normative competence (Block 1980). Accordingly, the problem of implementing legitimation increases with the recurrent crises of capital accumulation. Accumulation and legitimation, then, remain the conditions under which modern capitalism can expand. However, legitimation contradicts accumulation, as resources are withdrawn from the social players in charge of accumulation (capitalist class) to be utilized by the social players in charge of public administration (the officialdom of the state). Changes in favor of one or the other group of players could trigger crisis. A reduction in the economic solvency of the state could generate an inability to regulate the socioeconomic sphere and to overcome present elements of crisis. An increase in transfer from the economic elites to the state would signify a reduction of the potential for accumulation and a drainage of capital that would be detrimental for both groups in question. It follows that the state is called upon to mediate the contradiction between accumulation and legitimation, a task that the state is only partially able to manage.

The relative autonomy of the state is "relative" in that "the State is not independent of the structural contradictions and constraints of the capitalist economy," but the state does retain its own "autonomous political logic" (Jenkins and Brents 1989, 3). Therefore, the state cannot be viewed simply as an instrument of the capitalist class. For relative autonomy theorists, the state is more an arena for the mediation of class interests. An example of the relative autonomy approach in the area of sociology of agriculture and food is provided by the analysis of the land reform program carried out in Italy in the immediate post–World War II period (Bonanno 1988). In this case, the Italian state mediated between the interests of the urban bourgeoisie, who wanted to eliminate the power of backward rural landlords, and the aspirations of landless peasants who struggled to obtain land. The relative autonomy dimension is reflected in the class character of state action. Although the amount of land redistributed was insufficient to create a significant number of viable farms, it achieved the objectives of controlling the struggle of landless peasants and of providing a reservoir of labor for the urban bourgeoisie. The latter was directing a fast-growing process of economic expansion that was characterized by the gradual employment of unskilled labor from rural areas. Through the implementation of a limited land reform program, the state created the conditions for a steady and controlled flow of labor from rural to urban areas. Simultaneously, however, it allowed the creation of some viable farms and the possibility for rural labor to find temporary employment before their permanent migration to urban areas. In other words, the state accommodated the interests of both the urban bourgeoisie and the landless peasants, yet its ultimate effort was to create conditions favorable for capital accumulation to be carried out by the local bourgeoisie.

Jessop notes that the commonality of the instrumental and relative autonomy approaches is that they refute a "narrow institutional definition" of the state in which the " 'relative autonomy' of the state becomes total" and the state becomes a rarified "thing" in isolation from other institutions (1982, 22). That which the Marxist instrumentalists and relative autonomists refute, the rarified "thing," is what the state-centered Weberians assert.

The Autonomy or State-Centered View

State-centered theorists use a Weberian framework to contend that the state has its own interests and agenda apart from the ruling capitalist class, but that the degree of autonomy of the polity can vary in significance (Hooks 1990; Levine 1987, 1988; Levine and Lembcke 1987; Orloff and Skocpol 1984; Skocpol 1979, 1985; Skocpol and Amenta 1986; Skocpol and Finegold 1982). This approach emphasizes the politicized workings of the state in developing and implementing new policies. Skocpol and Finegold (1982) used the case of the agricultural policies of the New Deal to assert that the state acts autonomously by implementing policies contrary to the wishes of the capitalist class. Prechel (1990) uses the steel industry in World War II to show that the state chose production targets and forced expansion of production against the wishes of the steel manufacturers, supporting the idea of the state's autonomy.

The state as an autonomous institution has its theoretical base in the work of Weber, who indicates that there is separation between the state, the economy, and society. The Weberian view includes the observation that authority in modern societies is characterized by exclusive legal domination. This domination is achieved through the development of an institutionalized bureaucracy, which becomes literally the instrument of the contemporary state (Badie and Birnbaum 1983). For Weber, the more important question regarding the state was what it does versus the Marxist approach based on what it is.

What the modern state does, according to Weber, is monopolize the use of legitimate force within a given territory (1983). It is in the emergence of this form that the State becomes a distinct institution within society, the rarified "thing" referred to by Jessop (1983). According to Weber, it doesn't matter whether the state exists in a capitalist or a socialist society, because power in an institutionalized state ultimately rests in the hands of the top management of the nationalized or socialized enterprise (Held 1983).

Skocpol (1979, 1985) is the most prominent follower of the Weberian view of the state, arguing that "virtually all neo-Marxist writers on the state have retained deeply embedded, society-centered assumptions" (1985, 2). She asserts that instead of society itself, or a particular class, structuring the goals of the state, states may structure their own activities. "States conceived as organizations claiming control over territories and people may formulate and pursue goals that are not simply reflective of the demands or interests of social groups, classes, or society. This is what is usually meant by 'state autonomy' " (1985, 9). Weber's concepts of force, territoriality, and the dominance of institutional elites remain central to Skocpol's conceptualization of the state.

The Mixed Approach

The aforementioned theories have generally been employed in exclusive terms. An analysis explicitly rejecting the separation between the instrumentalist and relative autonomy positions is provided by Friedland (1983, 1991b). Friedland assumes that the role of the state in society is not given, but rather depends upon specific historical circumstances or the autonomous posture of the state in society. Employing the cases of various agricultural commodities, he demonstrates that the state is simultaneously called upon to organize various interests of dominant capitalist elites and to mediate between capitalists' interests and opposing interests emerging from other classes (an example of relative autonomy theory). However, he further demonstrates that in specific instances the state also operates as an instrument of the capitalists, as the latter directly and effectively control the action of the former (an example of instrumentalist theory). Empirically, he concludes, neither theory is sufficient to describe the complex patterns of state involvement in society. Paradoxically, each theory becomes correct under different historical circumstances.

Several students of the state have underscored the fact that the various theories of the state contain a number of important similarities (Bonanno 1987a; Carnoy 1984; Green 1987; O'Connor 1973, 1974, 1986). Among these are the overall tenets that economic growth and capital accumulation are not possible without the aid of the state and that the state cannot exist without the continuous existence of an accumulation process. In more specific terms this signifies first that accumulation of capital, economic growth, and the established position of the capitalist class in society depend upon the ability of the state to maintain the conditions necessary for the reproduction of capital. Second, accumulation of capital must be legitimized, and the state provides legitimation through the mediation of the various interests in society. This phenomenon refers to both mediation among various fractions of capital and among capitalists and other groups or classes. Third, the state obtains its financial resources from the taxation of revenue generated through the accumulation process. Accordingly, the state is dependent on the continuous existence of the accumulation process through economic growth. In essence, for all the above-mentioned schools of thought there is an intrinsic relationship between the process of capitalist development and the existence of the state apparatus (Bonanno 1991, 18).

An additional similarity shared by all approaches is the nation-centered conceptualization of the state. These approaches assume that the state and the economy are bounded within the same spatial context represented by the nation-state. It is within this context that the state can aid and regulate the development of the economy, and the economy can provide the basic financial resources for the continuous existence of the state.

Several recent studies that have carried out empirical analyses adopting a nation-centered concept of the state exemplify this approach. Hooks (1990) examines the U.S. Department of Agriculture (USDA) and its actions within the agricultural sector in several different time periods to find support for a state-centered theory. His work defines the state within the U.S. political apparatus. Devine (1985) employs a similar understanding of the state to examine social investment and social consumption outlays in the United States to support a political economy perspective. Quadagno (1990) looks at the Social Security Act of 1935 to support the theory of the U.S. state's relative autonomy in operating within the national sphere. Within the sociology of agriculture and food, Gilbert and Howe (1991) attempt to demonstrate the superiority of the relative autonomy theory in relation to the Agricultural Adjustment Administration's socioeconomic impact on farmers, again within a national construct. Finally, Green's (1987) examination of the structural changes within the national flue-cured tobacco industry posits that capital had gained power from the state.

Transnational Theories of the State

The debate on the role of the state has been redefined by scholarship underscoring the dissolution of the Fordist regime and the emergence of a transnational system, i.e., global post-Fordism (Bonanno et al. 1990; Constance and Heffernan 1991; Constance et al. 1993; Friedland 1991b; Friedmann and McMichael 1989; Heffernan 1990; Heffernan and Constance 1994). For example, Friedmann and McMichael (1989) examine the development and characteristics of the "second food regime," which mandates the establishment of transnational social relations. The first food regime was based on the emergence of strong national economies that governed the development of their nation-states. The European metropole organized extraction of cheap raw materials (beef, pork, wheat, corn) in their colonies and associated settler nations. These agricultural goods were then traded for manufactured goods from the core country. The second food regime emerged along with the transnationalization of the agriculture and food system. Transnationalization implies (1) the intensification of agricultural specialization and integration into agro-food chains that are characterized by economic concentration at both input and market sectors, and (2) a shift from food production designed as final use to one designed as industrial inputs for manufactured foods (Friedmann and McMichael 1989, 105).

Another example from Heffernan and Constance (Heffernan 1990; Constance and Heffernan 1991; Heffernan and Constance 1994) highlights the emergence of a few dominant food TNCs that globally source the world for profitable production sites and access to major markets. Their work emphasizes how these relatively few TNCs increasingly control the global food system and how this control is maintained by bypassing nation-states' regulations that increase the cost of business. They stress that "turnkey" transferability of modern production systems to almost any part of the world makes restrictive policies a risky venture for nation-states. For them, "the growing question is whether any country or set of countries can truly control their own food system in light of the shifting power relationships in the emerging global system" (Constance and Heffernan 1991, 136).

Bonanno et al. (1990) discuss the global dimension of agricultural and food production and the growing inability of the nation-state to successfully mediate the contradictions between accumulation and legitimation. These authors argue that, although it may appear that capital would prefer a laissez-faire economic system to the point of eliminating state intervention, the state performs necessary functions for capital accumulation, especially the legitimation function that sustains the cultural hegemonic project in support of capitalism (a la Gramsci). Although there might be some advantages to eliminating state intervention, the disadvantages that would emerge would be much greater and unacceptable to progressive and regressive social forces alike (Bonanno et al. 1990, 240–44).

Friedland contends that the emergence of TNCs suggests that the nation-state can only partially control these new economic forms "because much of their productive, manufacturing, distributing, and marketing functions are nationally-dispersed" (1991b, 49). Friedland asserts that internationalized capital would prefer to have some form of transnational-state to perform the vital functions of the nation-state at the global level. Although there are several embryonic attempts toward the creation of a transnational-state, no legal or political entity has yet emerged.

This literature stresses the inadequacy of the nation-centered concept of the state by arguing that, although the locus of action of capitalist accumulation has expanded to the global level, the locus of action of the state has remained within national boundaries (Bonanno 1992; Friedman and McMichael 1989; McMichael 1991; McMichael and Myhre 1991; Pitelis

1991; Picciotto 1991; Sanderson 1986). This fracture between the locus of action of the state and the economy requires a reconceptualization of the role of the state in the post-Fordist transition. There seems to be some general level of agreement that different forms of transnational-states are emerging, but that these forms are still nascent and not codified.

The recent debate on the relationship between the transnationalization of the economic sphere and the state underscores the effects that changes in the former have generated for the latter. In previous socioeconomic phases, i.e., the national and multinational phases (Bonanno 1987b, 1991, 21–24), the primary roles of the state were those in support of capital accumulation and social legitimation (O'Connor 1973, 1986). The manners in which the state performs actions aimed at fostering accumulation and legitimation have been called into question now that the accumulation of capital has entered its transnational or global phase.

Bonanno et al. (1994) argue that the changing role of the state in global post-Fordism requires important specifications. For example, it has been argued recently, particularly in the specialized media (e.g., *Business Week* 1990), that globalization has initiated a trend toward the weakening and eventual disappearance of the state as an element of coordination and regulation of economic activities. This position has been based upon arguments first discussed by scholars who view TNCs as more capable of coordinating the allocation of resources worldwide in terms of both market failure (i.e., the inability of market mechanisms to avoid socioeconomic contradictions) and inefficiency of the state apparatus (Bullock 1991; Kindleberger 1986). The thesis of the weakening and disappearance of the state has also been proposed by writers within the political economy camp (Ross and Trachte 1990). The action of TNCs has been viewed as "rendering the nation-state obsolete as the basic unit of economic decision-making for capitalist accumulation" (Borrego 1981, 727).

Despite the importance of these arguments, caution should be exercised in regard to the conclusions reached. Recent research has pointed out that the organizing, coordinating, and legitimizing roles performed by the state are still fundamental for overall economic growth and for the interests of the transnational corporations (Gamble 1988; Koc 1991; Pitelis 1991, 143). Accordingly, there is no conclusive evidence to support the existence of trends toward the disappearance of the nation-state.

It has been pointed out that the globalization of the economic sphere has affected the internal composition of the nation-state through the weakening of some of its apparati and through the strengthening of others. More specifically, sectors of the state oriented toward social programs have been weakened, while those oriented toward finance have been strengthened (Koc 1991; McMichael and Myhre 1991; Murray 1991; Pitelis 1991). Although the nation-state continues to perform vital functions for capital, the increased globalization of production has led to subordination of nationally oriented state ministries such as industry, labor, and planning to other ministries such as finance and prime ministers, which "are key points in the adjustment of domestic to international economic policy" (Cox 1981, 146). In this respect, the weakening of the state is only relative, as it refers to the role of the State in support of groups who benefit from the existence of social programs.

Four general positions exemplify the various understandings of the role of the state in the transnational sector (Bonanno et al. 1994). The first position interprets the state in terms of class control, a position exemplified by the work of William H. Friedland (1991a, 1991b). He maintains that the food policy is a critical national concern, because "unless there is a constant flow of food in abundance and relatively cheaply to the urbanized and industrialized populations of the advanced capitalist countries, domestic unrest will increase" (1991b, 51). Therefore, agriculture and food issues are fertile grounds for research on the transformation of the nation-state and the emergence of different forms of transnational- or supranational-states. Friedland defines the state as a "superorganic or metasocial process existing only partially in formal organization, but also in informal arrangements, agreements, and understandings that are imminent in the activities of key structural elements of advanced capitalist countries" (1991b, 52).

Although there was a global economy before the 1960s, prior to that time the economies were strongly tied to their nation-states. The defeat of the United States in the Vietnam War marked the end of the historical period based on the dominance of the nation-state as a political-economic form (Borrego 1981). About the same time the world system changed from one based on national political-economic interests dominated by national bourgeoisie imposing imperialism on the world to a transnational system dominated by TNCs. TNCs that were once attached to their host countries (as MNCs) have become "less concerned with specific national interests, national markets, and/or internal organization which is nationally-based and more concerned with a global orientation" (Friedland 1991b, 52–53).

For Friedland, since TNCs' global orientation can only be partially controlled by any one nation-state, "it follows that some new form of State form must emerge to function as the 'executive committee' of a new transnational bourgeoise" (1991b, 53). Nation-states can only control TNCs within their own territorial jurisdictions. Therefore, TNCs are currently unregulatable by any existing state form, even though TNCs would prefer to minimize uncertainty regarding regulation issues. For Friedland, it is the need to reduce uncertainty that necessitates the creation of the transnational-state. According to Friedland,

> While many aspects of production may have uncertainties, it is reasonable to expect that capitalists, in their relations with each other and, equally important, in their relations with the working class, will want to have some kind of predictability about the rules of the game. Means must be developed within which competition can occur, in which resources and markets can be developed, setting the boundaries on conflict so that capitalists will know what the costs of various production factors will be under any given set of conditions. It is because of the demand for these certainties, predictabilities, rules, that we can logically infer (once we have made assumptions about the character of the State as a given) that a transnational State must become emergent as transnational political economy spreads. (1991b, 53)

Friedland has emphasized that transnationalization of the economy requires the emergence of a state apparatus that can assert its powers beyond the boundaries of the nation. Historically, the state has emerged to minimize uncertainties in the accumulation of capital and to create a climate of business confidence. This process must be continued in the new transnational scenario if accumulation is to continue without unbearable contradictions. More specifically, given the continuation of conflict among various fractions of capitals (particularly between domestic capital and transnational capital) and between the interests of capital in general and subordinate social groups (environmental movements, consumer movements, ethnic and minority groups, organized labor, and so on), the organizational, mediative, and legitimizing roles must be performed in the new global scenario. Following Hechter and Brustein's (1980) historical analysis of the emergence of the capitalist state, Friedland argues that the emergence of the transnational-state must be linked to the issue of control of social opposition. In fact, it was through the action of controlling opposition that the domestic state emerged in the early phase of the expansion of capitalism. The primitive nation-state was created at the behest of feudal lords, who were struggling with the burghers and wanted protection for their property, rights, and prestige. "This State, in turn, would be captured by the bourgeoisie in the revolutionary struggles that marked the transition to modern capitalist States in Great Britain and France" (Friedland 1991b, 55).

The question of the state, then, is recast in terms of what opposition is developing at the global level that, in turn, will shape the terms of the emergence of a transnational-state. Friedland predicts that it will be the opposition of national capitals to international capitals that will characterize the social control function of the transnational-state. Following this line, he argues that some systematic approach to distinguish national bourgeois from international bourgeois is needed. Friedland presents the Trilateral Commission as an example of the "apparatus in which a normative organization of the transnational bourgeoisie and the transnational State" could emerge (1991b, 56) as a manifestation of the ensuing "transnationalization of the State" (Gill 1990, 1). Because the leadership of the Trilateral Commission contains a large number of prominent TNC executives, "the commission represents a potential venue for the development of the normative system of the transnational State" (Friedland 1991b, 56). For Friedland, it is the arena where the standards for product quality and other rules of the game are set that will illuminate the source of the new transnational-state.

Despite the availability of a number of possible outcomes, Friedland concludes that the path for the emergence of a transnational-state is unclear at best, though some embryonic forms of a transnational-state can be detected in the regulatory attempts of international organizations in the agricultural and food sector, such as FAO and OECD. He argues that the EC and NAFTA are not examples of transnational-states because they are regional collections of nation-states with no international state powers.

The second position maintains that the domestic state has already been transformed into a transnational-state by economic forces. This posture is exemplified by the work of Philip McMichael (1991) and his associates (McMichael and Myhre 1991), who argue that most current conceptions of the accumulation crisis and related restructuring of global capitalism incorrectly retain an unproblematic conception of the state. He asserts that the tendency to see the state as only a national construct is an untenable assumption. More specifically, the nation-state "is a recent, and increasingly fragile, form of state political organization" (1991, 72).

McMichael warns us to not conflate the state with the nation-state. Indeed the nation-state is a historical form derived from the social movements of the nineteenth century and it "is the state's national organization that is currently is crisis" (1991, 72). The nation-state is in crisis because capital is restructuring "above and below the administrative level of the nation-state" and therefore "directly challenging the national constraints (such as social and financial regulatory policies) that evolved within the social-Keynesian era" (1991, 72). Although capital is challenging the power of the nation-state, the nation-state itself is still critical to the movement of capital.

According to McMichael, "what we are witnessing today is the denouement of the nineteenth-century national movement" that "is arguably the principal source of tension in contemporary global political-economy" (1991, 73). Just as the "nationally-organized" capitalism of the nineteenth century has been steadily replaced by "transnationally-organized" capitalism in the post–World War II era, so has the relationship of the nation-state and the economy been transformed. McMichael provides examples of the changing role of the nation-state as the first international food regime was transformed into the second international food regime, discussed earlier by Friedmann and McMichael (1989). Basically, the first regime was characterized by an extensive form of accumulation that extended the industrialization of agriculture into the colonial and settler territories to source cheap, raw food materials for the European metropole. The second regime was characterized by an intensive accumulation process designed to strengthen the industrialization of agriculture by turning food products into industrial inputs for highly processed foods. Agricultural products became links in an agro-food chain that produced durable instead of perishable products.

McMichael maintains that the nation-state is "not so much a self-evident (national) economic unit, as a social and institutional complex governing segments of value circuits, some of which are international" (1991, 75). The domestic economy consists of fractions of capital that have different and often competing interests; the state is the locus of mediation for these conflicts. "The vortex within which the state currently finds itself is the increased tension between national and international forces" (1991, 76).

The breakdown of Bretton Woods and the centralization of banking capital on a world scale have influenced the globalization of production in such a way that national accumulation strategies have become problematic. New social movements pressure their governments for food security and other progressive actions, while international capital and its financial agencies, such as the IMF, impose austerity programs on Third World countries that reorient production from social issues to exports. In the vacuum of Bretton Woods "emerged new institutional mechanisms of control and cooperation in the global movement of capital," such as OECD, the World Bank, and IMF (McMichael 1991, 77).

McMichael criticizes the regulationist literature that argues that global arrangements stem out of negotiations among nation-states (Aglietta 1979; Gordon 1988). His critique is centered on the facts that, by assuming a nation-centered system, regulationists are unable to understand the recently established global mechanism of regulation and reify the concept of the nation-state. Conversely, he argues that the transnationalization of the economic sphere has already transformed the nation-state into a transnational-state that performs global functions of regulation. For McMichael, the transformation of the nation-state and the state system is a response to the current crisis of global economy. Capital is in the process of overcoming the constraints of national economic organization, where it has been subordinated to global commodity markets that cross national boundaries. The historical foundations of the nation-state, such as the institutions of central banking, currency exchange rates, national accounting, and international trade, are eroding (McMichael and Myhre 1991).

This transformation of the state is based on the integration of the nation-state into capital circuits that are increasingly transnational. First, the nation-state is faced with diminishing control of the activities of transnational financial capital (TFC) structures. For McMichael, TFC is the "anchor" of a new, globally constructed regime of accumulation. TFC is replacing national wage relations protected by welfare-states with a global wage relation organized by global capital. This decreased control, in turn, affects the existence of the nation-state by generating within it a shift of power in favor of finance ministries and to the detriment of program-oriented ministries. In other words, there is an effective uncoupling of the nation-state from banking capital and control of international monetary relations. Second, the establishment of a global agro-food system has diminished the state's capacity to control the composition of local food and agricultural production. The latter is increasingly extroverted and oriented toward the production of inputs for livestock feed and processed foods for affluent markets. The net result of the situation is that the state is no longer a "political mediator" between global capital and national bourgeoisie and the working class. Rather, it has assumed the role of facilitator of the requirements of global capital and the imposition of the global wage relation.

One example of how the nation-state is transformed in the new global regime is the structural adjustment loans (SALs) imposed by IMF and the World Bank on developing countries. These loans require the reorganization of domestic economies toward exports and away from social programs. In other words, SALs exemplify an erosion of the autonomy of the nation-state, especially in developing countries, as First World bankers and bureaucrats determine what national policy will be in the Third World. This process implies a loss of national control over the composition of local agriculture (McMichael and Myhre 1991). In conclusion, McMichael argues that the disorganization of the nation-state resulting from its transformation under TFC leaves some niches open for local action. For example, it may be possible that small farmers, local capitalists, and laborers might mobilize to create local and subnational agro-food systems that fill the gaps left by a food system organizing globally. The success of such efforts depends on a clear view of the part of these players regarding the supersession of the nation-state by the transnational-state and their abilities to bring their protests and proposals to the attention of both national policymakers and transnational institutions and firms. As one example he points to the recent successes of several native tribes and popular movements to protect the Amazonian rain forest (McMichael and Myhre 1991).

The third position emphasizes the limitations to the emergence of a transnational-state by underscoring the importance that the domestic state still retains in terms of processes of accumulation of capital and social legitimation. Koc (1994) states that globalization refers to both a process and a discourse. He maintains that globalization as a process refers to the "intensification of world wide social relations which link distant localities in such a way that local happenings are shaped by events occurring many miles away" (Giddens 1990, 64). Globalization as a discourse refers to the regulationist analyses of the transition from Fordism to post-Fordism.

Globalization is not a new process but rather an ongoing historical project that expands commodity relations over the globe. Although globalization is not necessarily new, the agents of globalization change over history, as "the conditions of accumulation have been continuously redefined and renegotiated by various socio-political actors at the national or international levels" (Koc 1991, 2).

The nation-state system originally brought together different regions and localities into one nation governed by one code of law. More recently, the nation-state has played the role of "universal regulator" through the introduction of standardization (such as time and the calendar). The nation-state system produced global mechanisms of regulation such as GATT and IMF to coordinate international trade. The global accumulation crisis and rise to power of TNCs have presented problems for the nation-state system, as the reorganization of global production by TNCs has produced limits to the sovereignty of nation-states (Koc 1991, 17).

Although the globalization of the economy has limited the ability of the nation-state to perform its historical roles, Koc (1991) maintains that the nation-state plays fundamental roles that are still unresolved transnationally. First, at the legitimative level the domestic state has historically been an agent that has homogenized and controlled ethnically, religiously, and politically diverse groups. These groups have been brought together through processes of legitimation that have culminated in the establishment of the "nation" and the ideological and normative dimensions associated with it. In this respect, the identification of diverse social groups with a country has allowed the creation of more or less cohesive social formations. The existence of an entity that could mediate and eventually resolve the differences among these groups at the international level is embryonic and certainly does not have the same powers of the domestic state.

Second, and at the level of accumulation of capital, the domestic state has been able to organize and maintain conditions amenable to capital accumulation within its territory by controlling labor and protecting capital. The globalization of production has altered these conditions so that protectionist measures are called for by various segments of capital while others advocate state action in opposing directions. In this respect, the ability of the domestic state to perform actions in protection of some segments of capital has been somewhat eroded. As far as labor is concerned, the transnationalization of production processes has not been accompanied by the transnational mobility of labor. Despite some partial exceptions such as the EC, the regulation of fluxes of labor at the transnational level is unresolved at best and could become the source of a resurgence of nationalist and racist sentiments among the working class.

The fourth position focuses on the concept of contradictory convergence and the resulting relationship between TNCs, the state, and subordinate classes in the transnational phase (Bonanno 1991, 1992). Bonanno argues that the globalization of production makes the traditional theories of the nation-state inadequate for understanding the state's role in an international economy. He contends that the previous three approaches pay considerable attention to the attempts of TNCs to bypass state action and/or to direct state action toward the fulfillment of their interests but have missed an important component of the emerging transnational-state. Little discussion is provided on the interests that both TNCs and subordinate classes would have in the emergence of a transnational-state and the implications and contradictions that this process would entail for all groups. TNCs need a political mediator or facilitator at the transnational level to provide a business climate of accumulation and to mediate legitimation demands from subordinate classes and between capitalist class fractions. Subordinate classes need a transnational state to help regain many of their losses centered around health and safety for workers, consumer protection, and environmental regulations that the transnationalization of the economy has accomplished. Some cases in point are the Amazon Basin/rain-forest issue, global warming concerns, the Circle of Poison problems, and challenges arising from the decline of unions.

According to Bonanno (1988, 1990), global economic restructuring, accomplished primarily through shifting of production across national borders, reliance on low-wage labor, and concentration of capital, has severely limited the nation-state's actions to protect the social and economic gains obtained by subordinate classes in previous periods. Bonanno argues that ongoing attempts such as EC, NAFTA, and OECD will only partially resolve the dislocation between the arena of nation-state power and the arena of capital accumulation, since these are not global entities but rather regional or "common interest" organizations. He concludes that the present "transnational-State vacuum" represents both a danger and an opportunity for progressive players. Although the lack of a clearly defined transnational-state "entity could provide the opportunity for creating a system in which equitable and democratic ends could be established," this situation could also be "transformed into an increasingly repressive global system" (1991).

Bonanno (1992) maintains that the emergence of supranational organizations such as EC and NAFTA is an early attempt at forming transnational states. He sees the strengthening of supranational organizations as the most advanced effort to establish political and legal elements of the states beyond national boundaries. In particular, EC, with its thirty years of history of the movement toward economic and political unification, is the "most advanced and sophisticated case" of a transnational state (1993, 15).

The debate on the emergence of the transnational state has recently been continued through a series of publications by many of the authors discussed above (e.g., Bonanno et al. 1994; Bonanno and Constance 1993; Friedland 1994; Llambí and Gouveia 1994; Marsden and Arce 1994; Moreira 1994; Schaeffer 1994). They underscore the historical characteristics of the transnational-state as well as its conceptual dimension through an analysis of the distinction between the state and the nation-state. They argue that, although it was possible in the Fordist era to equate the state with the nation-state, in global post-Fordism this equation can no longer be maintained. The creation of supranational states, e.g., EC, NAFTA, Economic Market of the South (MERCOSUR), the delegation of nation-state roles to nonnational entities such as supranational states and organizations of the United Nations like the World Industrial Patent Organization (WIPO) and Codex Alimentarius, the bypassing of state action by economic players, and other similar phenomena indicate that simply equating the state with the nation-state will capture only a limited dimension of state action. If this is the case, they argue, the operational definition of the state employed by studies that assumed a national center concept of the state must be modified.

The issue of the modification of the operational definition of the state can be addressed by identifying the roles of the state in capitalism and by tracing their evolution in global post-Fordism. By illustrating this process, it is possible, these authors maintain, to identify the limits of action of the nation-state and the evolution of such roles above and beyond national spatial boundaries. Accordingly they define the state in terms of "the set of apparatuses, agents and agencies which carry out the roles of the State at various levels and with various qualities" (Bonanno et al. 1994, 15). Following Friedland (1994), the notion of quality refers to the formal and nonformal channels through which state roles can be carried out. More specifically, they refer to the fact that these roles can be carried out through either recognized "formal" channels or alternative "non-formal" channels. It must be noted that the term "non-formal" does not refer exclusively to informal channels but is a more inclusive term of which informal activities are just a part. The authors also identify four spatial levels in the context of which these roles are carried out. In global post-Fordism these levels consist of the national, the international, the supernational, and the transnational. To be sure, imperatives, qualities, quantities, and levels are not static elements. On the contrary, their existence is to be verified empirically. In other words, they should not be treated as axioms, but rather as aspects which can be exhausted and/or maintained in the process of evolution of the socio-economic system.

Following the above definition of the states, these authors maintain that the reconceptualization of the action of the state must include the following:

1. The state can no longer be conceptualized exclusively in national terms, since the imperatives of the state have been delegated and/or subsumed by greater than national entities.
2. Conversely, any reconceptualization of the state must entail an international dimension. More specifically, state action has to be verified through the inclusion of more than national elements. This involves international, supernational, and transnational activities yet must not mandate the exclusion of the national dimension; all four levels should be part of the analysis.
3. The state cannot be conceptualized exclusively in terms of formal public apparatuses, agents, and agencies. As documented by the group's research, the delegation of state imperatives to a set of new players mandates the consideration of alternative spheres.
4. Accordingly, the reconceptualization of the state in global post-Fordism must include nonpublic apparatuses, agents, and agen-

> cies. The concept of nonpublic should not be equated exclusively with private. Private entities should be considered as part of the nonpublic sphere but should not be automatically equated with it. In this context, nonpublic has a broader meaning that includes quasi- or semipublic entities as well as quasi- or semiprivate apparatuses, agents, and agencies and all the various possible combinations that can be found in the continuum between "private" and "public."

One of the preconditions for the reconceptualization of the action of the state, these authors continue, lies in its contextualization within global post-Fordism. This requirement can be difficult to operationalize if the fact that global post-Fordism is characterized by a complex set of phenomena and contradictory trends is taken into account. Accordingly, they insist, it is necessary to create a framework that synthesizes global post-Fordism without diminishing its complexity and its dynamic character. For this purpose, global post-Fordism can be conceptualized as an ideal type characterized by a set of dialectical relationships. In this case the authors speak of the relationships of opposing forces that symbolize the evolution of this phase of capitalism and identify four dialectical relationships: deregulation/reregulation; fragmentation/coordination; mobility/embeddedness; and empowerment/disempowerment.

Deregulation/Reregulation

Strategies such as global sourcing and the implementation of neoliberal economic policies have allowed observers to argue the deregulation of socioeconomic processes. More specifically, it has been claimed that the state is either unable to regulate or has diminished or withdrawn from regulatory activities. The diminished intervention of the state has been paralleled, however, by two related phenomena. First, withdrawal of state intervention in some spheres, particularly the social sphere, has been accompanied by strengthening of intervention in other spheres. Accordingly, the withdrawal of state action has been developing only partially and contingently. Second, the space left from the withdrawal of traditional forms of state intervention has been claimed by other regulatory processes that involve the action of either public or nonpublic entities. In essence, global post-Fordism involves both deregulatory processes and, simultaneously, new forms of reregulation.

Fragmentation/Coordination

Among the most important characteristics of global post-Fordism are the spatial dispersion of production processes and the differentiation of consumption. In general, these processes have been associated with the concept of fragmentation. Post-Fordist fragmentation has been conceptualized as a strong departure from the Fordist concentration of production and consumption. These authors indicate, however, that the fragmentation of economic processes requires sophisticated systems of highly specialized coordination, as decentralized production processes need to be coordinated at a number of levels, for instance, at the level of control (e.g., control of the standards of production, control of quantity of output, and control of the quality of products). They must also be coordinated at the level of specialized markets. In this case, increased production of specialized commodities must be channeled to specific markets with very high levels of accuracy and rapidity. As illustrated by the case of the fresh fruit and vegetable sector, coordination between fragmented production and specialized niche markets is key for the continuous existence of the production system.

Finally, coordination is needed at the level of type of output. As perhaps best demonstrated by the cases of durable foods or computers, the various dispersedly produced components of each individual product must be generated in such a way that they can immediately be utilized in the creation of the final good. This situation refers to the cases of both ingredients and specific parts.

Mobility/Embeddedness

Global post-Fordist flexibility is synonymous with the rapidity of mobility of capital. The mobility of capital is paralleled, however, by relatively less mobile productive structures and much less mobile labor. Although capital can be moved electronically, allocation of investments depends on a set of variables such as availability of land and other required natural resources, infrastructures, and structures that require a greater amount of time to assemble than the simple electronic transfer of capital. Reallocation of labor depends upon an even greater quantity of time-consuming and spatially constraining items. These items can include getting out of current housing situations, paying for moving expenses, finding alternative employment for significant others, and so on. Aside from the material logistics involved, numerous human realities keep labor tied to particular places. Places are often connected to personal and familial histories as well as social and affective networks that are not easily left behind.

In short, they conclude, the mobility of the various productive elements transcends the existence of spatial enclaves that are touched by global post-Fordism. In locales left behind by the hypermobility of global capital, the negative consequences are often most evident in terms of unemployment, underemployment, environmental degradation, and community decay. Simultaneously, localities that "receive" global capital also experience a host of consequences, some positive and some negative. Global capitalism thus entails a local dimension that certainly touches processes involving labor, production, consumption, and institutions.

Empowerment-Disempowerment

Global post-Fordism has been viewed by many as a strategy to weaken subordinate classes in society. The crisis of unions, the generation of "bad" jobs that are increasingly replacing well-paid employment, and the proliferation of part-time and flexible occupations are a few of the most significant cases in point. Similarly, capital hypermobility has been identified as one of the reasons for the economic decay of regions and communities. The severity of these phenomena has been so great that global post-Fordism has been interpreted as a totalizing phenomenon leaving little room, if any, for resistance. Although this scenario has a significant degree of validity, these authors indicate that along with the disempowering of some social groups, movements that resist global capital have emerged. In the sphere of agriculture and food, the growth of the environmental movement is certainly a significant example. Moreover, it has been argued that within global post-Fordism TNCs have reached unmatched levels of power. Their ability to bypass regulations created to protect other social groups attests to this claim. However, these authors also point out that TNCs are not exempt from experiencing significant contradictions and problems. Perhaps one of the most relevant of these is the limited level of intercorporation coordination generated by the restructuring of the state.

Regulationist Views of the State

The literature reviewed above indicates that there are a variety of reasons to argue for the continuous existence of the nation-state. Simultaneously, there are numerous reasons to argue that a form of transnational-state should also exist. These arguments are also found in positions developed in fields outside the sociology of agriculture and food (e.g., Borrego 1981; Fine and Harris 1979; Picciotto 1991; Pitelis 1991), which conclude that the collapse of the international order typical of the Fordist regime requires the emergence of new transnational forms of the state. However there is consensus that neither the empirical nor the theoretical form of the "new" state is clear as yet.

For example, Radice (1984) points out that the internationalization of the economy has gone so far that not only does no significant amount of national autonomy remain, but the economy has become so internationalized that the strategies of national autonomy would be extremely problematic. According to Poulantzas, internationalized capital has the unfair advantage of direct access to the various international state apparatuses such as IMF, the North Atlantic Treaty Organization (NATO), and OECD, which can be used externally to coerce nation-states (1974). For Poulantzas, nation-states themselves have no power of their own but instead have to crystallize the relative strength of classes and fractions within them (1975). From this view, the question of the nation-state versus TNC is a false one. Poulantzas's view provides some insight into why "nation-states may (not) support the emergence of transnational bodies; and/or change their attitudes on this issue in different time periods" (Pitelis 1991, 141). In other words, the nation-state can be politically captured by subordinate classes, probably temporarily as capitals' noncooperation with the state would soon precipitate a more capital-friendly state, who can then codify into law their progressive agenda.

Pitelis (1991) argues that TNCs and nation-states exhibit both rivalrous and collusive behaviors, depending on the particular historical conjuncture. More specifically, the degree of rivalry and collusion "will depend heavily on whether the relationship refers to TNCs' 'own States,' or 'host

States," as well as whether the states in question are 'strong' or 'weak,' [developing countries] DCs, or LDCs" (Pitelis 1991, 142). Pitelis presents Hymer's (1971, 1976) view to support his point. Hymer observed that the increasing bargaining power of TNCs asymmetrically threatens the autonomy of all nation-states. More specifically, TNCs threaten the LDCs' "weak States" more than the DCs' "strong States." Hymer agrees with Marx that TNCs assist the development of LDCs but do so in a dependent and uneven fashion. "TNCs he suggested shape the world to their image, creating 'superior' and 'inferior' states" (Pitelis 1991, 141).

Pitelis points out that TNCs need both the nation-state and international organizations on the grounds that, domestically, they need to control and exploit labor through the existence of some form of local state. Internationally, capital needs international organizational forms, particularly in relation to the crisis of U.S. hegemony worldwide. According to Pitelis: "The very fact that TNCs increase their power over labour and the nation-state due to their relative flexibility of operations, suggests that TNCs need a system of nation-states . . . if they are to exploit this advantage! Also, in order to play-off nation-states against each other" (1991, 143).

Again relying on Hymer (1971), Pitelis argues that transnationalization of the economy implies that TNCs now need the support of nation-states other than their own. He asserts that this does not lead to the supersession of the nation-state by the TNC or by international organizations. Rather, the nation-state's particular advantage is its legitimation role, "in terms of exploitation and/or creation of nationalism" (Pitelis 1991, 144). In other words, capital needs the nation-state, especially the strong ones.

Furthermore, capital also needs international organizations to surrogate the functions of the nation-state at the international level. Pitelis presents the work of Fine and Harris (1979) to argue that transnational organizations better serve transnational capital. They assert that the nation-state is determined by the relations between all classes within it, not just capital. Accordingly, class conflict that results in the electoral defeat of capital's representatives constrains the nation-state's ability to act instrumentally on the behalf of capital. Fine and Harris observe that the distinct advantage of international state–like apparatuses is that by distancing themselves from the point of class struggle, such organizations are better able to pursue the interests of capital, in particular its fractions representing international capital (Pitelis 1991, 145).

Pitelis also argues that nation-states need transnational capital, which provides investment and economic growth. Hymer and Resnick (1971) observed that the relationship between capital and the nation-state was generally a symbiotic one until transnationalization started to threaten the state's autonomy. Transnationalized capital's demands to make investment more attractive in protectionist nation-states "threatened to expose the class nature of the state, thus diminishing its legitimizing ability" and its ability to represent the interests of capital accumulation in general and to maintain a stable system (Pitelis 1991, 145). When this ability is threatened, the possibility of rivalry between nation-states and "their" TNCs increases.

In conclusion, Pitelis argues that TNCs need international organizations in order to

> globalize the socialisation of the costs of its global operations, which however does not lead to a withering of the nation-state. Instead, the nation-state persists alongside such organizations for reasons related to benefits for transnational capital, but also for labour and interests of the state functionaries. (1991, 146)

In general the developing nation-states tend to assist the requirements of transnational capital for the emergence of transnational organizations in the expectation that the strong state will share in the benefits of its TNCs abroad. An example of Pitelis's concept of collusion, this collusive alliance is threatened by the increasing power of TNCs to "maximize their private benefits, without sharing the benefits with some nation-states, or even at their expense" (1991, 146). This leads to his concept of rivalry.

Pitelis maintains that "labour movements should try to mobilise support towards pressing the nation-state to reduce the mobility of TNCs, and increase their commitments to their 'host' (including home) countries labour" (1991, 147). He argues that LDCs should should try to obtain control over the flow of information and technology so that they can develop some form of independent industrialization. Finally, Pitelis warns us that the induced divisions within labor itself, such as employed versus unemployed, DCs versus LDCs, racism, and sexism, combined with the increasing spatial international division of labor keep progress restricted to fairly narrow venues. Nonetheless, although "this increases the grip of transnational capital on labour and on the system as a whole, . . . one should not despair, but equally there are few reasons for 'revolutionary optimism' today" (1991, 147).

Picciotto argues that a "major feature of the political economy of the current period is the existence of acute tensions in the international state system and struggles over its reformulation" (1991, 44). Although the crisis of U.S. hegemony has opened up the possibility of a new international order, "the continued strength of the U.S. allows it to dominate the terms of the transition" (1991, 45). He states that the analyses of nation-states and of transnational organizations have focused too finely on the former and not the latter. He argues that there has been a "mushroom growth of international organizations" since 1900 that has created "the formal structure of a network comprising innumerable meetings and contacts of officials, managers, and representatives of all kinds" and that these nongovernmental organizations provide a "minimum degree of coordination of State regulation necessary to permit the international reproduction of capital" (1991, 52). Although these seemingly disparate organizations do not by themselves resemble a "world State" or "world government," "the very multiplicity and heterogeneity of the international bodies that have grown up, and their primarily ideological function, are the key to their importance" (1991, 53).

Picciotto maintains that the increasing international mobility of capital fostered some attempts to develop some international coordination of financial regulation that have so far been inadequate. These attempts to tighten up the international regulatory system have entailed "increased bureaucratic-administrative coordination and corporatist bargaining at the international level," which have "created an increasing problem of legitimation" (1991, 56). Although traditionally the technical-bureaucratic characteristic of most international regulatory processes relied on the the nation-state to provide legitimation, increasingly the nation-state is seen by new social and class movements as ineffectual in countering the globalization of capital and production, resulting in the "renewed internationalisation of social and class movements" seeking legitimation at the international level (Picciotto 1991, 57).

For example, in response to increasing concerns over the growing power of TNCs, in the late 1960s the United Nations developed an international Code of Conduct for TNCs and also set up a United Nations group of experts to produce a report on multinational enterprises (TNCs). In 1972 the International Chamber of Commerce, "sensitive to the needs of international business for legitimacy," drew up the first Guidelines for International Investment and Multinational Enterprise (Picciotto 1991, 57). For Picciotto, the United Nations Code of Conduct and other similar business guidelines are "largely symbolic, a reaction to and an attempt to control the growing criticisms of and actions against TNCs from the late 1960s" (1991, 57).

Picciotto argues that the labor movement needs to work with new social movement (NSM) groups to build a new internationalism that can combine the strengths of bureaucratized and disciplined class politics with the flexibility of grass-roots popular social movements. This coalition must then "attempt to capture and transcend the forms of international state which have developed as parts of the internationalization of capital" (1991, 59). Examples of some successes of such a group are the infant formula issue, the pesticide circle of poison issue, and the South African divestment issue.

In conclusion, Picciotto reminds socialist activists that human emancipation is not possible just by capturing the nation-state and subsequent efforts to further it along the road to socialism. Indeed, the relative success at these efforts in the forms of social-democratic and welfare-states has been countered by the internationalization of capital and production, which bypasses these territorially distinct jurisdictions of democratically created citizen, worker, and environmental protection. Finally, a new view of the state can "develop more sophisticated analyses of the contradictions of the state and the ways they can be exploited to build the strength of popular movements" (1991, 60).

The changing role of the nation-state in global post-Fordism is also discussed by Scott Lash and John Urry, David Harvey, and Bob Jessop. According to Lash and Urry (1987), the growth of the global corporation limits the abilities of individual nation-states to "organize" their national economies. As Crouch and Wheelright argue, the major reason is that

> the increasing trade is between corporations and not nations, and is being conducted on the basis of the comparative advantage created and manipulated by the very corporations that reap the benefits. The flows of trade, their direction, volume and pricing, are more and more at the discretion of global corporations, which make these

> administrative decisions internally for the purpose of maximizing global profit. (quoted in Babcock 1984, 154)

The growth of internal trade greatly increases the vulnerability of individual nation-states that are unable to control the flows of globalized trade "taking place under the visible hand of the global corporation" (Lash and Urry 1987, 197). Nation-states are also becoming more dependent on the export earnings to counteract import penetration, imports controlled by the global corporations. The limited ability of nation-states to regulate and direct current flows of international trade is illustrated by the rapidly growing U.S. debt, which makes it impossible for even a strong nation-state like the United States to successfully organize its national economy (Lash and Urry 1987).

According to Lash and Urry,

> There are a number of causes of these major developments: The breakup during the 1950s and 1960s of the old empires and "spheres of influence" that were very much part of "organized capitalism"; the development of computerized systems of information, storage and retrieval, and of satellite-based communications that have enabled different operations within global corporations to be located in different parts of the world; the growth of new international money markets (particularly the Eurodollar and Eurobond Market), which enabled large companies to engage in financial speculation and indeed to become, in a sense, bankers as they took advantage of differential exchange and interest rates; the general acceleration in the movement of capital, which has become less government-regulated as new loci of economic and financial power developed towards the end of the post-war boom; and the reduced ability of states to control interest and exchange rates and hence the tendency for economies to be subjected to balance of payment crises. (1987, 198)

The deregulation and decentralization of finance capital and the effects of substantial privatization of international money have undermined the capacity of individual nation-states to pursue national economic policies. The demise of Bretton Woods, the dropping of the gold standard for United States dollars, American corporate investment and technology transfer to the nonindustrialized world, and the shift in the balance of profit from industrial capital to commercial and financial capital indicate the dramatic disorganization of nationalist capitalist economies in the last twenty years. Industrial and financial capital have both been internationalized "but with separate and unco-ordinated circuits," which has "massively weakened the individual nation-state which places its economy within one or other vicious circles and makes the state unable to regulate and orchestrate its national currency" (Lash and Urry 1987, 208–9).

Lash and Urry present Aglietta's scenario as a possible outcome of the unstable international credit system:

> A fully fledged international credit system, deterritorialized and beyond regulation by any sovereign state. Yet these three essential functions of money are neither unified under a supranational monetary body nor managed through inter-state cooperation based upon universally agreed principles. Indeed, the instability of the reserve function leads to rivalry among convertible currencies which destabilizes the complementary relations joining national regimes of growth. (Aglietta 1982, 25)

The fragmented character and growing success of political activity by new social movements facilitate the decline of corporatist stability associated with organized capitalism. Neocorporatism is on the decline as working-class capacities decline and big labor can no longer make long-term deals with big capital and the big state to stabilize and regulate economies. These corporatist class compromises mediated by the welfare-state were fundamentally national agreements regarding resource distribution. Import penetration and dependence have fostered a new international division of labor, which deemphasizes mass production in Western welfare-states. Without the mass-production labor process, big labor cannot represent enough workers to provide corporatist solutions.

During the same time that national economies were penetrated by global corporations, the public sector was growing rapidly. The fiscal and accumulation crises of the Western welfare states brought on austerity programs whereby social resources were diverted into export industries. The disorganization associated with a lack of a corporatist compromise brought Keynesian stabilization strategies under attack from both labor and capital. This took the form of the "great mobilization" of labor in the late 1960s and early 1970s, while for capital,

> The disorganizing processes of labour-market fragmentation and internationalization laid open the possibility to pursue different, anti-Keynesian and anti-corporatist, strategies. The growth in public-sector employment and the concomitant formation of powerful unions created a potential crucial division within the labour movement. Internationalization, and especially import-penetration, exacerbated this division. (Lash and Urry 1987, 235)

In their conclusion regarding the demise of corporatism, Lash and Urry state that

> corporatism assumes the existence of three national corporate bodies: capital, labour, and state. Disorganized capitalist internationalization—with its new prominence of the export of means of production, of finance capital and most importantly of commodities—has been crucial in the undoing of neo-corporatism. With internationalization, capital above all ceases to be a national class, the state loses control over important economic processes, and the heterogeneity of working-class interest groups is severely aggravated. (1987, 280)

Although working-class capacities are in decline,

> we are not pushing a thesis of the imminent demise of the proletariat. We think that the industrial struggles of the late 1960s and 1980s are evidence of continued existence of considerable working-class capacities. The struggles of the late 1960s were, however, importantly infused with the (non-class specific) radical-democratic discourse which also informed the contemporary social movements; and those of the mid-1980s were often part of a bid to pre-empt the attempt of large capital to bid adieu to the proletariat. (Lash and Urry 1987, 13)

To summarize, the post-Fordist transition and resulting global restructuring have been determined by the operation of three parallel processes. First, the nation-state has been subjected to (1) a number of internationalizing effects "from above," including the emergence of global corporations creating a new international division of labor and the related growth of new circuits of money and banking separate from those of national industries that "are literally out of the control of individual national economic policies"; (2) a variety of decentralizing processes "from below," including the disintegration of the "spatial fixes" of industry, classes and cities characterized by the decentralization of industry and population, the declining attractiveness of mass organizations and mass production, and the declining importance of class interests; and (3) the growth of the service class from "within," which has led to an emergence of service-class issues in politics, on professionalization strategies, and on "cross-cutting forms of social division and cultural conflict" (Lash and Urry 1987, 300–301).

Regarding the U.S. experience, the development of organized capitalism at the turn of the century was characterized by an "exceptional fusion of banks and industry" aided by a state that was "the instrument of the economically dominant class" (Lash and Urry 1987, 305). The influence of progressivist ideology in the early 1900s "contributed to the manner in which the state developed a degree of autonomy during the 1930s New Deal; and it contributed more generally to the development of the American 'welfare-state' " (1987, 305). For these authors, it is only in the U.S. case that the rise of the service class's, not labor's, progressivist agenda generated "such a large measure of the challenge to capital and the creation of some degree of relative autonomy of the state" (1987, 305). As the service class developed itself, it transformed American society by weakening labor's agenda, strengthening American capital, and expanding the growth of an educationally based stratification system.

Along with the demise of working-class capacities, Lash and Urry argue that it is increasingly questionable whether the modern welfare-state can survive. The fragmentation of support for welfare policies combined with the fragmentation of the production processes coordinated by powerful global corporations create situations whereby "national economies are increasingly out of control" (1987, 309). The globalization of banking, such as the Eurodollar and Eurobond markets that are "essentially privatized and outside the control of either national governments or of international regulatory bodies such as the IMF, and emerging dependence of all national economies on world trade destabilizes nationalist welfare-states (1987, 309–10).

According to Harvey (1990), the demise of Bretton Woods in 1971 and the ensuing adoption of the flexible exchange system signaled that the

United States no longer had the ability to control global fiscal and monetary policy. Since the demise of Bretton Woods, "all nation states have been at the mercy of financial disciplining" (1990, 165). Two examples of the "simple acknowledgement of external financing power over internal politics" are the French Socialist government's "turnaround in the face of strong capital flight after 1981" and the British Labour government's concession to "austerity measures dictated by the International Monetary Fund in order to gain access to credit" (1900, 165). Although there had always been a "delicate balance between financial and state powers under capitalism, . . . the breakdown of Fordism-Keynesianism evidently meant a shift towards the empowerment of finance capital vis-a-vis the nation state" (1990, 165).

The rapid reduction in costs of transportation and communication have to a large degree released many firms from being geographically anchored to a particular national or regional space because of dependence on raw materials or markets. This increased geographic mobility has resulted in the often mentioned "new international division of labor" characterized by the "proliferating mechanism of co-ordination both within transnational corporations as well as between different sectoral commodity and product markets" and the rise of powerful NICs in the "locational re-shuffle of the world's industrial production" (Harvey 1990, 165).

During this period, U.S. dependence on foreign trade increased twofold and imports from developing countries increased tenfold. The United States changed from a creditor to the world's largest debtor nation and lost the lead in global banking. At the same time neoconservative regimes took over the governments in the West, as the growth of welfare-state spending associated with entitlements combined with decreased state revenues to bring on a fiscal and legitimacy crisis for the state. Resulting decreases in state welfare expenditures negatively impacted labor and the social wage; even democratic and labor regimes were constrained to follow similar reductions in the welfare-state. According to Harvey,

> The gradual withdrawal of support for the welfare state, and the attack upon the real wage and organized union power, that began as an economic necessity in the crisis of 1973–75, were simply turned by the neoconservatives into a governmental virtue. The image of

> strong governments administering powerful doses of unpalatable medicine to restore the health of ailing economies became widespread. (1990, 168)

Increased international competition and slow growth forced all states to maintain an attractive business investment climate, which required that labor and other social movements had to be controlled. Therborn's (1984) comparative study of European states shows that austerity measures, fiscal retrenchment, and the erosion of the social compromise between big labor and big government were common to every state in the advanced capitalist world. Even though individual states still retain significant power to intervene in labor contracts, their "accumulation strategies" (Jessop 1982, 1983) have "become more strictly circumscribed" (Harvey 1990, 168).

Increasing instability in international financial markets has forced many states to become more interventionist. Using the U.S. Federal Reserve's power to alleviate the Mexican debt crisis of 1982 and the U.S. Treasury to broker the approximately $20-billion write-off of Mexican debt in 1987 are two examples of increased intervention in international markets. Other examples are the massive outlays of the U.S. Federal Deposit Insurance Corporation (FDIC) and the Federal Savings and Loan Insurance Corporation (FSLIC) to bail out the collapsing banking and thrift industries. After the October 1987 market crash, the New York Federal Reserve and the Bank of England intervened dramatically to stabilize their respective currencies. "The role of the state as a lender or operator of last resort has, evidently, become more rather than less crucial" (Harvey 1990, 169).

Harvey argues that the institutionalization of economic summits by the capitalist states in 1975 was, and still is, a collective attempt to win back some of their lost power. Their designation in 1982 of IMF and the World Bank as the "central authority for exercising the collective power of capitalist nation states over international financial negotiations" is an example of this attempt (1990, 170).

Although the conservative regimes criticized the welfare-state for its costly programs, the massive government deficits in the United States associated with defense spending provided "whatever economic growth there had been in world capitalism in the 1980s, suggesting that Keynesian practices are by no means dead" (Harvey 1990, 170). The neoconservative commitment to free-market competition and deregulation does not exactly fit with the waves of mergers and joint ventures between rivals during the 1980s. According to Harvey:

> Arenas of conflict between the nation state and trans-national capital have, however, opened up, undermining the easy accommodation between big capital and big government so typical of the Fordist era. The state is now in a much more problematic position. It is called upon to regulate the activities of corporate capital in the national interest at the same time as it is forced, also in the national interest, to create a "good business climate" to act as an inducement to trans-national and global finance capital, and to deter (by means other than exchange controls) capital flight to greener and more profitable pastures. (1990, 170)

Although the modalities, targets, and capacity of state intervention have changed during the transition, this does not mean that it has diminished. Particularly for the case of labor control, "state intervention is more crucial now than it ever was" (Harvey 1990, 170).

Regulationists vary greatly in their views of the state. According to even themselves, their accounts of the state are a major weakness of their framework (Boyer 1986b). Jessop (1990) provides his own interpretation of the different regulationist views of the state.

Regulationists who have a narrow economic focus, such as the law of profit, tend to characterize the state as an "ideal collective capitalist." For example, the Groupe de recherche sur la regulation d'economies capitalistes (GRREC) notes that the state apparatus and budget are deeply linked to the laws of profit and that the Fordist state regulates growth consciously, while the liberal state leaves regulation to the invisible hand of competition (Rosier and Dockes 1983, 204–5). From this view, priority is always given to economic function over state form (Jessop 1990, 196).

French regulationists like Aglietta who focus on the codification of the wage relation argue that the state penetrates civil society and becomes a constitutive element of the wage relation. The state is the forum where the various structural forms of regulation are concentrated and their coordination is managed over time (Aglietta 1979, 32, 383). For Boyer, "the state often plays the determining role in the diffusion, and sometimes even the genesis, of the essential institutional forms" of the wage relation

(1986b, 53). Jessop argues that this view suffers from reductionist and functionalist dangers. More specifically, it is reductionism to assert that the state's "essential role is to manage the tensions and contradictions in regulation" and functionalism to "claim that the State must to do so for accumulation to proceed" (Jessop 1990, 197).

Several French, West German, American radical, and Amsterdam regulationists see the state as the institutional embodiment of a class compromise that extends well beyond the wage relation. For example, Delorme and Andre (1982) show how French public expenditures reflect and entrench specific class compromises. West German theorists focus on the state's role in coalescing a hegemonic power bloc that underwrites specific accumulation strategies and societalization forms. The American radicals view the state as involved in all four accords—capital-labor, citizen, international, and domestic competition—and also stress the fact that the democratic state is the site of conflict between the logic of capital and the logic of citizenship (Bowles and Gintes 1982). The Amsterdam school stresses the state's centrality as the locus of comprehensive control domestically, as well as either a proponent or opponent of transnational penetration (Holman 1987–1988; van de Pijl 1988, 1989). Although all of these approaches stress "class struggle and the hegemonic role of one or another fraction of capital, the State is none the less accorded a key role in constituting and managing the struggle" (Jessop 1990, 198).

Some regulationist theorists have concentrated on the regulation of global regimes and are especially concerned with the issues of international cohesion. Such studies have focused on the diffusion of Fordism, multinational corporations, trade relations, and the emerging complementarities among different modes of growth. Others have focused on the cohesion of the international regime and the role of dominant or hegemonic states in defining and managing that regime. For example, the Amsterdam school has emphasized the "role of a transnational bourgeoisie in shaping the international order" (Jessop 1990, 198).

Another factor that influences how various regulationists view the state is their respective use of different, substantive theoretical frameworks. Regulationists inspired by Keynesianism tend to focus on the issues of postwar growth and see the state's role as an instrument of economic management. For analyses that focus on economic and political business cycles, class struggle is seen as a struggle over distribution issues in which the state's role is to manage the balance of forces that provide for matched production and consumption. Marxist analyses linked to state monopoly capitalism tend toward "functionalist capital logic and/or reductionist class-theoretical analyses of the State" (Jessop 1990, 199). Analyses in the Gramscian and West German traditions provide a more state-centered approach, in which the balance of forces is "overdetermined by State forms and the sui generis dynamic of the political region" (Jessop 1990, 199).

Finally, for most regulationists the state's role changes according to the specific historical period. The students of Kondratieff's long waves argue that the state's activities have expanded as the liberal, competitive stage of capitalism was replaced by social-democratic polities and economic monopolies. This view can be seen in studies of state monopoly capitalism as well as studies of the intensive regime, with its associated Fordist monopoly mode of regulation or SSA. Some authors argue that the transnationlization of the economy has also been related to the shift in the state's role. For example, Overbeek (1988) argues that the state's role is more interventionist during crises of a mode of regulation than crises within the mode. Accordingly, crises within a mode are managed by established routine adjustments of the economy, while crises of the mode require the state to perform a more interventionist role to restructure and solidify a new mode of regulation before it "once again withdraws somewhat to forms of intervention consistent with the new mode" (Jessop 1983, 199).

Although the state can never be absent from the institutions that underpin the modes of regulation, Jessop criticizes the regulationists for generally subsuming the role of the state under a general account of the structural forms associated with the mode of regulation. With few exceptions, regulationists have borrowed previous accounts of the state and attached them to their new approach to economics. "They have not really applied the same approach to the state itself nor have they tried to integrate more adequate state theories" (1990, 199).

For Jessop, if the regulationists' approach is valid for the study of the economy, then it should also apply to the polity. Jessop sees the state neither as an ideal collective capitalist performing instrumental functions nor as simply an agent of pluralist forces. For him it is better to see the state as an ensemble of structural forms, institutions, and organizations whose functions for capital are deeply problematic. The state not only manages a complex mix of instruments and policies to secure the conditions for capital accumulation and maintain an unstable class compromise, it must also struggle continuously "to build consensus and back it with coercion" (1990, 200). According to Jessop, the state itself can be seen as a complex ensemble of institutions, networks, procedures, modes of calculation, and norms as well as their associated patterns of strategic conduct" (1990, 200). Therefore, the state cannot be seen as "regulatory machine" that is lowered into place as capital needs it. Rather, the state is both an object as well as an agent of regulation. The particular way the formal aspects of regulation are ordered by contributing agents is what gives the political sphere its "relative autonomy" and means one cannot treat politics just as "concentrated economics" (Jessop 1990).

According to Jessop, contrary to an instrumental view of the state and the mode of regulation, the regulationist works of Delorme and Andre (1982), Delorme (1987), and Baulant (1988) reveal that "much recent work on regulation has emphasized how the fragmented structure of the state affects its capacities to engage in economic management or crisis-resolution and, conversely, how its "sui generis" dynamic and the structural legacy of institutionalized compromise mean that it has a certain inertial force" (Jessop 1990, 200). Jessop states that a regulationist approach to the state would treat it like the wage relation or the commodity, as something that needs regulation itself. He provides four crucial premises for a regulationist theory of the state (1990, 200–201). First, there are problems involved in managing the state itself as a crucial site of regulation. The state does not preexist as a fully constituted, internally coherent, organizationally pure, and operationally closed system but rather as an emergent, contradictory, hybrid, and relatively open system. Therefore, there can be no inherent unity between the state and the institutional ensemble, and the unity that the state provides for the mode of regulation is always relative and must be created within the state system itself through specific activities and means of coordination. The state is also held responsible for promoting the interests of the "illusory community" that exists in the wider society, a society of which the state is only a part. Regulationists have studied this responsibility in terms of the various class compromises or accords that define the "illusory community" whose interests and social cohesion must be maintained by the state. Without a great consensual hegemonic project, the measure of internal unity and resulting social base that the state needs to secure the political conditions to support the new regime of accumulation is absent.

Second, the state's strategic options are defined by the form of political class domination inscribed within a given state system. This refers to the specific configuration of state branches, apparatuses, and institutions, their specific powers and prerogatives of action, their specific relative autonomies and institutional unities, and their specific patterns of domination or subordination. These aspects overdetermine the specific aspects of the capitalist state and produce a specific system of structural and/or strategic selectivity. The historical specificity of the state means that it is not equally accessible to all social forces, cannot be controlled or resisted to the same extent by all strategies, and is not equally available for all purposes. Different political regimes inevitably favor the access of some forces, the conduct of some strategies, and the pursuit of some objectives over others. Accordingly, an essential feature of any stable mode of regulation is the structural and strategic selectivity inscribed within the political forms that correspond to it.

Third, the relative importance of the state's role varies with the object of regulation. Indeed, the state's activities are themselves a focus of struggle with a major impact on different modes of regulation. This is especially important because of the state's close ties to national economic space and means that the internationalization of the economy will require new forms of regulation. The best way to approach this set of problems is in terms of specific national modes of growth and the historic blocs with which they are associated. For example, Lipietz notes that a mode of development/growth is based on a coherent and stable combination of a technological paradigm, an accumulation, a regime, and a mode of regulation. Lipietz states that the "regime of accumulation would appear as the 'macroeconomic' result of the functioning of the mode of regulation, on the basis of a model of industrialisation" (Lipietz 1986, 14). In this triangular relation, the mode of regulation plays the role of a two-way mediating link between the technological paradigm (material base) and the regime of accumulation (hegemonic bloc). For Lipietz, the state played a central role in providing coherence in the Fordist triangular relationship. It is the state that materializes the "structural forms" which crystallize "institutionalized compromises." The state is both archetypal and also serves as the ultimate guarantor of the other structural forms (Lipietz 1985, 11).

Fourth, the regulationists' predisposition to see the state in structural terms regarding objects of regulation needs to be modified to incorporate the subjects of regulation. For example, which strategies are adopted by what social actors toward the state and state power in the struggle to restore, maintain, or transform a given mode of regulation? These questions cannot be meaningfully answered in purely abstract terms. Failing to specify agents and mechanisms leads to class reductionism, sterile analysis, and overemphasized structural factors. Historically, regulationist studies have focused on the structural relations between the state and economic categories and have shown little concern with how they are mediated in and through the strategic conduct and routine activities of social forces.

Jessop presents the "gestural references" to Bourdieu's concept of "habitus" by Lipietz (1986, 1988) as a modest attempt to bring agency into regulationism. Some of the West German regulationists have paid attention to agency in the form of the Gramscian notion of political parties. Political parties play a key role in mediating between the state, individuals, and institutions, acting as a clearinghouse for pluralistic demands of various interest groups and providing a vital legitimation function. Some Parisians have also argued that the mode of regulation is characterized by a balance of force between collective class players organized as political parties. As particular modes of regulation decay, these players "find it hard to represent new class interests and/or shape new regulatory forms" (Jessop 1990, 203). Therefore, for new modes of regulation to emerge, there must be the development of "new collective subjects" (e.g., Gramsci's "new collective wills") (Aglietta and Brender 1984, 21–22, 162–65, 209–10).

Noel also argues that political parties have a key role in producing the political realignments that consolidate new "implicit contracts" favorable to an emerging mode of regulation (1988, 19, 23–25). Jessop adds that the role of bureaucrats and other nonparty intellectuals has to be included in the development of norms and procedures that sustain a given mode of regulation. He presents Gramsci, who noted how hegemony could be grounded in the factory under Fordism.

There are many sites on which partial modes of regulation and specific regulatory procedures are mediated. Recently, writers have shown how NSMs act not only as relays and agents of crisis in a mode of regulation, but also can serve as useful venues of experimentation with possible new structural forms, norms, and strategies suitable for a new mode of regulation (Mayer 1987, 1988; Roth 1987). In conclusion, Jessop states that the mediation of regulation has to be examined through specific social practices and forces, and not through "speculative" structuralist categories (1990, 204).

4

THE TUNA FISHING INDUSTRY IN THE FORDIST PERIOD

"Porpoise!" The lookout spots a large school near the horizon. The tuna boat slows and stops, and two fast outboard skiffs, or pongas, are lowered into the water. The drivers, in life jackets and crash helmets, strap themselves into their seats and check their radio headsets. They give a signal, and the boat gets under way, the pongas following close astern. The school is much closer now, and the skipper scans it with powerful binoculars, searching for signs of tuna. He sees jumping fish, and the real action begins now. The mastman in the crow's nest signals the chasers and they surge forward across the tops of the swells. He directs them by radio as they flank and harry the porpoise school, trying to tighten it and slow it down. The mastman calls, "Stand by!" and the deck crew gets ready to release the net.

The chasers are running ahead of the school now, and the porpoise are beginning to mill in a tight circle. The mastman shouts, "Let her go, let her go!" and a big skiff drops into the water from its perch atop the net pile on the fantail, pulling the end of the net with it. The boat circles the porpoise school at full speed, and the huge stack of webbing on the net platform melts swiftly as the half-mile-long seine plays out.

The mastman gives directions to the helmsman on the bridge and at the same time coordinates the movements of the two chaser pongas, trying to keep the school together until the set is completed. The entire school, a thousand porpoise or more, has been encircled.

> As the boat completes the circle, the other end of the net is picked up from the big skiff, and pursing begins. The purseline, acting as a huge drawstring, gathers up the bottom of the seine and in a few minutes the net is closed. Now begins the long, hard job of hauling in and emptying the net.
>
> What sort of fishing is this? (Perrin 1968, 166)

The discussion of the previous two chapters was intended to provide a background for the analysis of the tuna-dolphin controversy. Framed in the process of transition from Fordism to global post-Fordism, this case provides us with information that can shed light on the complex of changes taking place in contemporary society. In this respect, the following chapters should not be viewed so much in terms of a test of the accuracy of the theories presented above. Rather, they should be read as the presentation and analysis of a case that illustrates significant aspects of the socioeconomic restructuring of late capitalism. It is this historical dimension that takes primacy over a theoretical posture centered on comparisons between various accounts of the transition to global post-Fordism. Evaluations of the adequacy of the theories presented above can and will be carried out, but the intent here is to present a case of globalization that indicates, albeit partially, alterations of established patterns of social interaction and the consequent redefinition of significant social relations. With this objective in mind, this chapter will briefly discuss the development of the tuna fishing industry and its evolution until the early 1980s, the period when the first signs of the restructuring process emerged. The restructuring of the tuna fishing industry is then discussed in Chapter 5.

THE HISTORY OF TUNA FISHING

Stone Age people fished for tuna in dugout canoes with barbless hooks made from horn. Japanese records reveal that metal hooks were introduced in the eighth century and that by the twelfth century nets were used to catch tuna. By the nineteenth century, tuna fisherman used poles and hooks with live bait. Since remote times, the fishermen from the Azores and Madeira have used the method of chumming tuna with live bait and catching them with poles and lines (Ben-Yami 1980). Portuguese fisherman from the Azores brought this tuna fishing method to California, where it developed into an important industry. Technological advances in ship design and mechanized fishing facilitated the expansion of the Eastern Tropical Pacific (ETP) tuna fishery over wider areas (Ben-Yami 1980).

The development of the U.S. tuna canning industry can be traced to 1903, when a San Pedro, California, sardine packer switched to canning albacore ("white-meat") tuna when sardine supplies were scarce. Consumers' acceptance of the canned tuna supported the development of fishing fleets in the San Diego and San Pedro areas. Increased public demand for tuna expanded the tuna fishery to include skipjack and yellowfin tuna, the "light-meat" tuna. Tuna processing facilities were concentrated on Terminal Island in the Long Beach–Los Angeles harbor, and San Diego was the main base of the U.S. tuna fleet (Rockland 1978). By 1930, about four-fifths of the U.S. tuna harvest was caught off the coasts of Mexico and Central and South America (USITC 1986, 7).

Prior to 1957, tuna fleets consisted almost entirely of baitboats using poles and lines with live bait. The anchovies used as bait were caught along the coasts near the tuna fishing grounds. This technology was relatively labor intensive; bait fish had to be caught, hooks had to be baited, and several fishermen were stationed on each boat to catch tuna. In 1957, two technological innovations revolutionized the tuna fishing industry (McNeeley 1961). The advent of nylon fishnetting and the introduction of the power-operated Puretic hauling block facilitated the development of the "purse-seine" fishing method, which freed tuna fishermen from their dependence on coastal bait fisheries. Purse seining had been unsuccessfully experimented with through the early 1900s, but available technologies proved to be cost-inefficient and therefore inadequate. "The capacity to operate far offshore led vessels to areas where yellowfin tuna were more abundant and where they associated more frequently with dolphins than they did near the coast" (Hofman 1981).

Between 1957 and 1961 many of the tuna baitboats underwent conversion to purse seiners. These conversions markedly increased vessel productivity by increasing the distance and reducing the length of fishing voyages. New tuna boats that were large, fast, and specifically designed as purse seiners began to be built. By 1969, "the realization of huge profits to be made in tuna fishing led to a rapid expansion of the fleet" (Rockland 1978, 5). From 1969 to 1974 the total capacity of the San Diego fleet almost doubled (Rockland 1978), and by 1974 there were over 7,000 tuna fishermen operating more than 2,000 tuna vessels in the United States. About 75 percent of the U.S. tuna catch was accounted for by 150 purse seiners operating primarily in the Eastern Tropical Pacific (ETP) and to a lesser extent in the Atlantic, off the coast of Africa (Salia and Norton 1974). By 1985, purse seiners accounted for 99 percent of the U.S. tuna fleet's capacity (USITC 1986, 8).

The five major commercial species of tuna are divided into temperate and tropical groups (Joseph 1973). The tropical tuna species are yellowfin, skipjack, and bigeye; temperate species are the bluefin and albacore. Tuna are classified as pelagic fish, which means that they migrate extensively across the open seas. In 1974 there were about forty nations involved in commercial tuna fishing; of these Japan, the United States, Taiwan, Korea, France, and Spain (in rank order) accounted for over 80 percent of the catch of the six major species (Joseph 1973, 1972). Japan accounted for about twice as much (40 percent) of the global catch as did the United States (468,000 metric tons versus 239,000 metric tons). Still, Japan's percentage of the tuna pack had declined steadily from 1962 (Asada 1972), even though the Japanese tuna fleet of 1,200 vessels in 1974, mostly long-liners, was the largest and most modern in the world (Salia and Norton 1974).

Since ancient times there has existed a fishery for tuna in the western Pacific, and skipjack tuna has been an important part of the Japanese diet for many years. Japan developed fishing bases for longline and baitboat operations in the Pacific Islands region during the 1930s. By 1938 there were more than 7,600 Japanese tuna fishermen operating in South Pacific waters (Doulman 1987). World War II drastically reduced Japan's fishing industry, and as part of reconstruction after the war, distant-water fishing was targeted as a growth sector, and legislation "paved the way for the virtually unrestricted movement of Japanese distant-water tuna fleets into the Pacific islands region" (Doulman 1987, 36).

From the end of World War II to about 1952, the Japanese baitboat and longline tuna fisheries and supporting bases were confined to the western and central Pacific Ocean. By the early 1960s the Japanese had established tuna bases in American Samoa, Fiji, French Polynesia, and Vanuatu. These bases were designed to supply tuna to Heinz (Star Kist) and Ralston Purina/Van Camp (Chicken of the Sea) canneries on American Samoa (Doulman 1987, 36). In the late 1950s and early 1960s, the

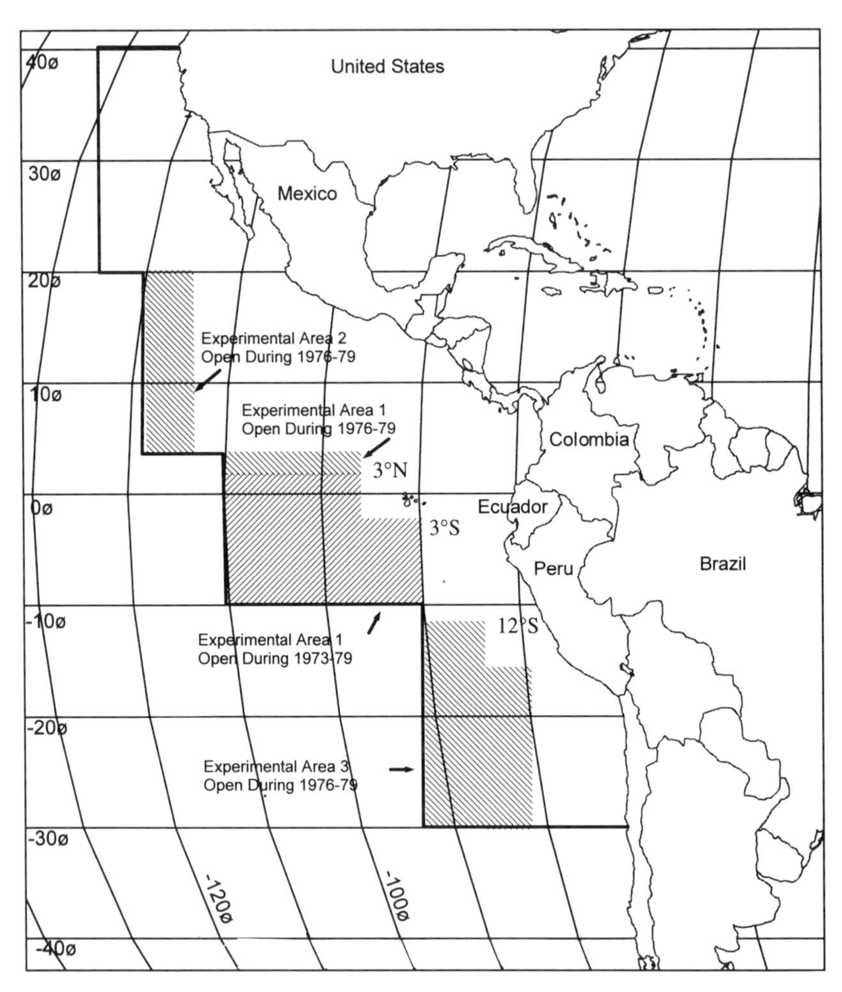

FIGURE 4.1. Eastern Tropical Pacific (ETP) and CYRA areas. *Source:* Adapted from Inter-American Tropical Tuna Commission, "The fishery for tunas in the Eastern Pacific Ocean" (1989).

Japanese longline fishery expanded into the ETP but has since declined (Salia and Norton 1974).

During the early 1970s the catch rates for the Japanese long-liners for certain species declined significantly, resulting in increasing costs per pound of production. The growth of the Japanese economy during this time also increased the costs of labor and fishing gear. In response to these economic constraints, the Japanese industry established bases in foreign countries and entered into agreements to access lower cost foreign labor and capital, particularly in Taiwan, Korea, and Malaysia (Moal 1972; Asada 1972). In spite of these efforts to reduce operating costs, the Japanese fleet experienced decreasing profits due to increasing competition from the "expanding fleets of nations such as Taiwan and the Republic of Korea, which among other things generally have lower labor costs." (Salia and Norton 1974, 12).

In 1974 the world's tuna fisheries could be divided into two major groups: the longline fisheries of Japan, Korea, and Taiwan and the surface fisheries based on pole and live bait or purse seines of the United States and Japan. Longline fisheries accounted for the highest percentage of the tuna catch, followed by baitboats and purse seiners (Salia and Norton 1974). Between 1960 and 1972 the capacity of the world tuna fleet doubled. By 1979 most tuna stocks except for skipjack tuna appeared to be "fully or almost fully exploited and some stocks may be overexploited" (Joseph and Greenough 1979, 11).

Longline fishing is designed for catching large fish at great depths and involves setting main lines that each contain branch lines with a total of up to 2,000 hooks (Salia and Norton 1974). Live bait and pole fishing is based nearer to shore than longline fisheries because of its dependence on bait fish, which are used to attract and concentrate the tuna. When schools of tuna are spotted, the fisherman use poles and lines to land as many tuna as possible before the school dissipates. "This type of fishery is considered to be extremely difficult and arduous work" (Salia and Norton 1974, 9).

Purse seining uses a large net of up to a mile long and 600 feet deep that is pulled by a skiff (a small boat with a powerful motor) to surround the tuna. The net is then drawn shut at the bottom to prevent the tuna from escaping. This technique is capital-intensive as compared with longline or baitboat fishing. By 1974, the purse seine, "a highly efficient method of surface fishing," was rapidly replacing bait fishery in some areas (Salia and Norton 1974, 9). According to Salia and Norton:

> The development of the purse seiner, along with restrictions on catch of yellowfin in much of the eastern Pacific, has resulted in many U.S. vessels entering the Atlantic tuna fishery. This tuna fleet is currently the most modern and far-ranging fishing fleet based in the United States. These modern vessels, however, represent a very

large capital investment and if increasing worldwide effort on tunas results in declining catch rates, these vessel owners will likely find it more and more difficult to maintain an adequate return on their investment. (1974, 13–14)

PORPOISE FISHING

For reasons uncertain to scientists or fishermen, in the ETP—a triangle that stretches from San Francisco to Hawaii to Peru—there exists a dolphin fishery where large yellowfin tuna swim under dolphins. "The bond linking tuna and spotted dolphins is remarkably strong. It may persist through much or all of the seining operation. During seining, tuna and dolphins continue to associate so tightly that to catch dolphins also means to catch tuna" (National Research Council 1992, 7). Fishermen use the dolphin to find the tuna, and therefore this method is called "porpoise fishing."[1] In 1972, between 50 and 70 percent of the annual U.S. yellowfin catch in the ETP came from schools associated with dolphins (unpublished data from Inter-American Tropical Tuna Commission [IATTC] cited in Coe and Sousa 1972). Tuna primarily associate with two species of dolphin, spotted dolphins called spotters and spinner dolphins called spinners. To a lesser degree tuna associate with the common dolphins, called whitebellies (Perrin 1968, 1969, 1970; Green, Perrin, and Petrich 1971). During the late 1950s San Diego fishermen developed porpoise fishing using purse-seine methods, which involves the use of a large net that encircles the tuna (and dolphins) and allows the capture of a great number of large yellowfin tuna in one setting of the net. Yellowfin tuna is most popular in the United States and represents about 10 percent of the global tuna catch (Levin 1989).

The purse-seine method was developed by San Diego fishermen in response to (1) reduced access to traditional fishing grounds, (2) high numbers of U.S. tuna boat seizures by foreign nations claiming territorial fishing rights, and (3) low-cost dumping of tuna in the United States by Japanese tuna fishermen (Tennesen 1989; Kraul 1990). Both points one and two were largely the outcomes of developing nations' expansion of their territorial waters to 200 miles, referred to as their exclusive economic zone or EEZ. The purse-seine method uses speedboats, helicopters, and small explosives called seal bombs to herd the dolphins into a net of

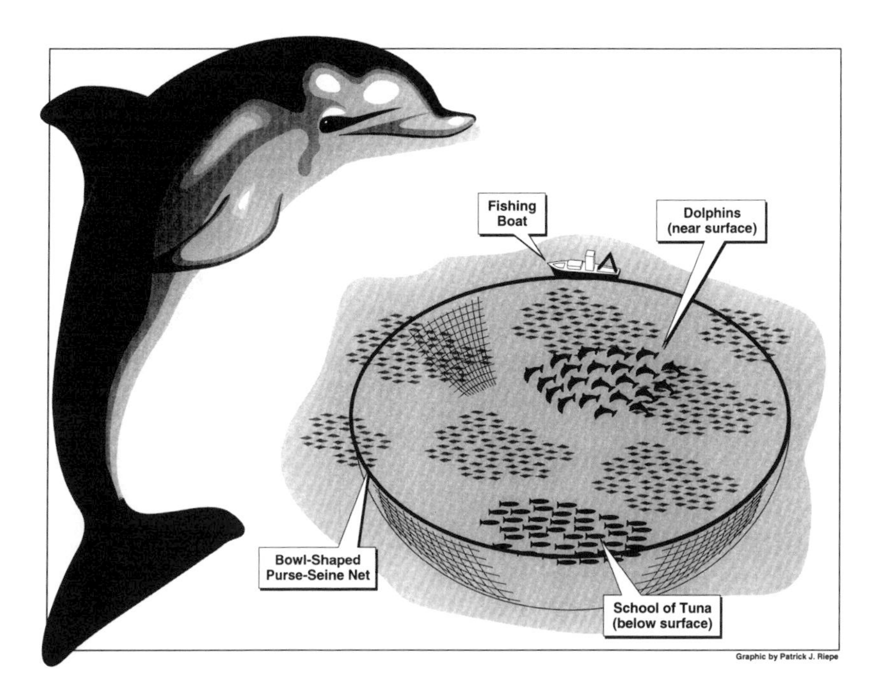

FIGURE 4.2. Purse-seine fishing method.

nearly a mile in circumference. The tuna stay under the dolphins, and although the dolphins could easily jump over the nets' corklines, they do not, and often get tangled in the net and drown.

The new technologies based on "setting on dolphins" facilitated enormous yellowfin tuna catches of up to 250 tons per set (Tennesen 1989) and an average of 18 tons per set by the 1980s (Brower 1989). By 1976, the U.S. tuna fleet had grown to 130 huge purse seiners that measured up to 200 feet in length, weighed 1,500 tons or more, and cost $10 million (Kraul 1990). The new technology provided very high levels of accumulation; the problem was that it systematically killed hundreds of thousands of dolphins a year. During the decade of the 1960s, about a quarter of a million dolphins were killed annually in the ETP (Brower 1989, 37). During both 1960 and 1961, shortly after the advance of purse-seine methods, around three-quarters of a million dolphins were killed (NMFS cited in Joseph and Greenough 1979). During 1970 and 1971 dolphin kills averaged 309,000 (*Marine Fisheries Review* 1979; see Table 4.1).

Table 4.1. Estimated Porpoise Kills for Principal Stocks by Vessels of All Flags in the Eastern Pacific Ocean, 1959–1976 (in thousands)

Year	Offshore Spotted	Eastern Spinner	Whitebelly Spinner	Total[a]
1959	68	40		109
1960	534	278		853
1961	446	232		713
1962	106	55		169
1963	133	69		213
1964	255	133		407
1965	297	155		475
1966	281	146		449
1967	195	101		311
1968	164	85		262
1969	331	158	14	529
1970	308	134	26	492
1971	185	99	31	315
1972	273	44	21	338
1973	120	44	30	194
1974	75	22	18	115
1975	106	26	39	171
1976	85	9	40	134

Source: National Marine Fisheries Service as cited in Joseph and Greenough 1979, 143.
[a]Totals include kills of other tuna species.

In the early days of the porpoise fishery, no successful technique existed to safely release the dolphins from the purse-seine net. Fishermen often went into the water to release the dolphins over the net's corkline as the net was drawn onto the boat, but by this time most of the dolphins had died from drowning. This activity was time-consuming and dangerous because of sharks. As a result, dolphin mortality was often very high (e.g., over one-half million per year). In an attempt to develop alternative dolphin release methods, ETP fishermen experimented with two techniques: the backdown procedure, followed ten years later by the Medina safety panel (Joseph and Greenough 1979, 141).

The backdown procedure was developed sometime around 1960. When approximately two-thirds of the purse seine had been brought aboard ship and the dolphins were at the far end of the net from the ship, the captain applied reverse power, causing the back of the corkline at the back of the net to be pulled under the water and pass below the dolphins, which were mostly at the surface. Too much backdown could permit tuna to also escape from the net. To prevent tuna loss, men in speedboats were stationed at the dolphin release area to pull the submerged corkline back to the surface if the school of tuna moved toward the back of the net. The men in the speedboats also worked to release dolphins caught in the net. This procedure was repeated until all or most of the dolphins were released (Joseph and Greenough 1979, 139,141).

The backdown method significantly reduced dolphin mortality, but still many dolphins died in the nets because of entanglement. In 1971 fishermen began to experiment with a smaller diameter webbing in the area of the net where dolphins concentrated. The Medina safety panel was named after its inventor, Captain Harold Medina. Since most of the dolphins' contact with the net occurred during backdown at the back of the net, fewer dolphins get entangled in the finer mesh net. According to Joseph and Greenough (1979, 142), "Because of its obvious benefits, the Medina or safety panel modification was rapidly incorporated into the gear of all United States vessels and many vessels fishing under other flags." The adoption of backdown procedures and Medina safety panels was stimulated by the enactment of the Marine Mammal Protection Act (MMPA) on October 21, 1972.

Marine Mammal Protection Act of 1972

In 1972, public outcry and indignation over the killing of dolphins associated with tuna fishing brought passage of MMPA, which mandated that over a period of time "commercial operators' marine mammal kills be 'reduced to insignificant levels approaching zero' " (Godges 1988, 24). At the time of the legislation, the U.S. tuna fleet was responsible for 85 percent of dolphin kills in the ETP (Godges 1988). MMPA banned the killing of marine mammals (whales, seals, sea lions, dolphins) but contained a two-year exception for commercial tuna fishing. The American Tunaboat Association, a trade association of U.S. fishermen, was accorded a general permit for the incidental killing of dolphins in the course of commercial fishing operations. The tuna industry was also supposed to develop new fishing methods less harmful to dolphins. "However, bowing to industry pressure, Congress granted the U.S. tuna fleet a two-year grace period to develop new procedures to minimize dolphin mortality. No new techniques were forthcoming, but in those two years 500,000 more dolphins died in U.S. tuna nets" (Davis 1988, 486).

MMPA also provided for import sanctions against other nations that did not protect dolphins. The law states: "It is unlawful to import into the United States any . . . fish, whether fresh, frozen, or otherwise prepared, if such fish was caught in a manner . . . proscribed for persons subject to the jurisdiction of the United States" (Joseph and Greenough 1979, 169). The director of the National Marine Fisheries Service (NMFS), a division of the U.S. Department of Commerce, was charged with identifying whether foreign fleets' tuna imports were "accomplished in a manner which does not result in incidental mortality and serious injury rate in excess of that which would have resulted from fishing under United States regulations" (Joseph and Greenough 1979, 169). The foreign fleets did not have to conform to the specific requirements of U.S. regulations, but they did have to have comparable dolphin mortality rates. For Joseph and Greenough, "any real resolution of the problem will require the cooperation of all coastal nations that border on the eastern Pacific as well as all nations whose flag vessels harvest tuna in the area" (1979, 168).

When the two-year extension expired in October 21, 1974, the NMFS issued new general permits through 1976 to the American Tunaboat Association on behalf of American ETP fishermen for the incidental killing of dolphins. To obtain the general permit the tuna industry was required to adopt specialized dolphin safety gear and maintain detailed records of their fishing activities. NMFS also was granted the authority to establish a kill quota for dolphin species and to prohibit "dolphin sets" on stocks for which kill quotas had been reached. Finally, as part of the law NMFS organized an observer program that placed observers on one-third of U.S. tuna boats to document the number of dolphin kills associated with tuna fishing (Joseph and Greenough 1979).

Joseph and Greenough (1979) report that the conflicting language in MMPA made implementation difficult. For example, Congress' findings stated that "certain species and population stocks of marine mammals are, or may be, in danger of extinction or depletion as a result of man's activities," and, as a matter of policy, "they should not be permitted to diminish below their optimum sustainable population(s)" (OSP) (Joseph and Greenough 1979, 144). In another section of MMPA the language was: "In any event it shall be the immediate goal that the incidental kill or incidental serious injury of marine mammals permitted in the course of commercial fishery operations be reduced to insignificant levels approaching zero mortality and serious injury rate" (Joseph and Greenough 1979, 145). These goals are conflictual: One mandate is to maintain stable dolphin populations, which does not require the elimination of purse-seine fishing in the ETP, while another mandate is to eliminate incidental dolphin mortality and injury, which requires that elimination. "As long as tuna are caught in association with porpoise by purse-seining methods, some mortality will occur" (Joseph and Greenough 1979, 154). These differing interpretations of MMPA produced a struggle between environmentalists and industry supporters that continued for twenty years.

Environmental Lawsuits and MMPA Amendments

Angered by NMFS's issuance of general permits and the resulting perceived lack of governmental enforcement of MMPA, environmental organizations accused NMFS of improperly issuing the permits and in 1974 brought suit against the Secretary of Commerce and other federal officials. After many months of litigation, Judge Charles R. Richey of the U.S. District Court, District of Columbia, on May 31, 1976, found that NMFS had "failed to fully discharge its obligations under the MMPA and ordered that the taking of porpoise should cease effective May 31, 1976. This ruling, which threatened to seriously cripple the United States tuna fishing industry, triggered a series of legal maneuvers which delayed the implementation of Judge Richey's order" (Joseph and Greenough 1979, 146).

Judge Richey's decision established a dolphin mortality quota that dropped from 78,000 in 1976 to 31,150 in 1979 and down to 20,500 in 1981. The 1976 quota was for 78,000 dolphins without reference to specific tuna stock and was granted on June 11, 1976. To establish the quota, NMFS held a workshop to quantify dolphin stock populations and evaluate their level of depletion; 1977 and later quotas used these calculations to establish species-specific quotas. The 1977 quota was 43,090 offshore spotted dolphin and 7,840 whitebelly spinner dolphins. No quota was granted for eastern spinner dolphins (Joseph and Greenough 1979, 152). According to Joseph and Greenough:

> Hence, unless a major breakthrough is made in fishing technology, it is clear that further quota reductions must, at some point, begin to

> limit tuna production. But if the tuna industry can succeed in meeting the 1979–1980 quotas, there will be no threat to the maintenance of large and viable porpoise stocks. Thus, further quota reductions, if they are to unattainably low levels, would seem difficult to justify, except on humanitarian grounds. (1979, 154)

The conclusions of the workshop were based on analysis of the OSP of each tuna species. Eastern spinner populations were found to be "either below or close to the lower end of OSP range," while offshore and whitebelly dolphin populations were "probably within" or "within" the range (Joseph and Greenough 1979, 151). "It seems probable that any losses resulting from an early cessation of setting on common dolphin schools could be made up by fishing on other types of porpoise schools" (Joseph and Greenough 1979, 154). Again, according to Joseph and Greenough:

> If the incidental kill must be reduced to an insignificant level approaching a zero mortality, then this would imply that the OSP and the MSP [maximal sustainable population] and a virgin stock must be maintained. But with existing technology, it is not possible to eliminate porpoise mortality without eliminating seining for porpoise-associated tuna and greatly reducing yellowfin yields. (1979, 148–49)

Joseph and Greenough stated that if the increasing political and legal pressure to curtail dolphin mortality resulted in the necessity to eliminate porpoise fishing in the ETP, that action would precipitate at least a 69 percent decreased catch in the area. This catch reduction would translate into "a loss of nearly 85 million dollars in dockside value or, very roughly, one-half billion dollars industry-wide" (in 1977 dollars; 1979, 25).

In 1980 the Reagan administration extended the 20,500 quota level indefinitely (Godges 1988), an action that ended the managed decline in dolphin kills. In 1984, MMPA was amended to extend this quota level indefinitely (Holland 1991). Instead of abolishing the intentional netting of dolphins, MMPA's quota system institutionalized the practice (Davis 1988). Under the Reagan administration, the U.S. tuna industry and supporting agencies were able to more effectively contest the 1972 law and thereby significantly weaken, and even alter, the intent and language of MMPA—"insignificant dolphin kill levels approaching zero" was redefined to be equal to 20,500 per year.

The Inter-American Tropical Tuna Commission

The conflict caused by increasing harvesting efforts by many nations in the ETP led to the formation and subsequent resource management activities of the IATTC (USITC 1986, 102), which is one of four major international organizations concerned with the management and scientific study of tuna resources in the ETP. As mentioned above, the ETP had been a major baitboat fishery for yellowfin and skipjack tuna since the end of World War II. IATTC was established in 1949 by a treaty between the United States and Costa Rica; by 1977, Canada, France, Japan, Mexico, Nicaragua, and Panama also had joined. IATTC has its own internationally recruited scientific staff. "The duties of the IATTC are to study tuna and other fish caught by tuna fishing vessels within its geographical area of responsibility and to recommend management measures designed to maintain stocks at levels that will produce maximum yields on a sustained basis" (Joseph and Greenough 1979, 14).

In 1966 IATTC research prompted the adoption of an overall catch quota for yellowfin tuna within a specified area of the ETP (see Figure 4.1) known as the commission's yellowfin regulatory area (CYRA). The annual quota of tuna taken in the CYRA was established on a first come, first served basis. This "race to harvest" resulted in overcapitalization in "superseiners" and a rapid reduction in the length of the fishing season in the CYRA; from ten months to three months (Salia and Norton 1974, 39). Although the quota system of 1966 adequately maintained tuna stocks in the CYRA, by 1979 other problems threatened its stability (Joseph and Greenough 1979).

In 1976 IATTC enlarged its mandate to deal with the growing tuna-dolphin controversy. Three basic objectives were adopted: (1) to maintain tuna production at a high level, (2) to maintain stocks of porpoises at or above levels that would ensure their survival in perpetuity, and (3) to make every reasonable effort to ensure that porpoises are not needlessly or carelessly killed in fishing operations. To attain these objectives a special program of activities was developed in 1977, which included the organization of an international seagoing technician program to facilitate the estimation of dolphin kills and dolphin stocks. The commission also set up programs to keep abreast of current developments in dolphin safety gear and dolphin and tuna behavior (Joseph and Greenough 1979, 170).

Joseph and Greenough (1979) present two main problems facing IATTC: issues regarding the distribution of allowable catch by nation and the rapid expansion of the global tuna fleet. Since the CYRA quota operates on a first come, first served basis, the largest fleets caught the largest share of the allowable catch. "For years the United States took nearly 100 percent of the catch, but, since the implementation of the yellowfin management program in 1966, their share of the combined yellowfin and skipjack catch has fallen to about 65 percent" (1979, 15). In the 1970s developing Latin American nations adjacent to the CYRA began building their tuna fishing fleets. These nations argued that they deserved special access to the tuna resources just off their coasts and that the "distant water fishing nations" (DWFN) should be restricted. In the early 1970s IATTC "initiated certain country quotas, primarily to allow for new entrants into the fishery and to protect certain of the fleets with smaller vessels that have no alternative fishing areas" (1979, 24).

Developed DWFN like the United States argued that tuna was a highly migratory fish and therefore was not the property of any particular nation, but belonged to whomever could catch them. In particular, the U.S. industry felt that it deserved considerable access to the CYRA tuna resources since it was responsible for the technology, expertise, and investment that developed the ETP tuna fishery (Joseph and Greenough 1979).

At the same time that developing nations were beginning to make their claims to national fishing waters, the global tuna fleet was expanding rapidly. The consistently rising demand for tuna during the 1960s and 1970s outstripped supply. The resulting "anticipation of profits has resulted in intense competition among countries and among industry elements within countries for the available catch" (Joseph and Greenough 1979, 15). From 1960 to 1977 the capacity of the international surface fishing fleet in the ETP increased from 44,000 short tons to 169,000 short tons; the U.S. share of that catch decreased from 90 percent to 73 percent over the same period. This represented a 280 percent increase in fleet capacity as opposed to an 80 percent increase in catch in the ETP over the same period. The decreased catch pushed many of the ETP purse seiners to move to fisheries adjacent to the CYRA or to the Western Tropical Pacific (WTP) or the Atlantic fisheries (Joseph and Greenough 1979).

In 1979 Joseph and Greenough noted that unless substantial new resources of tuna could be identified and harvested or unless tuna prices rose substantially, overcapitalization would result in "vast economic disruption in the fishery including, with virtual certainty, many bankruptcies" (1979, 32). In response to the growing economic uncertainties related to overcapitalization, beginning in the early 1970s "many major United States tuna-boat owners, foreseeing economic difficulties in the future, have sold their vessels to processors who seem to be integrating their industry vertically" (1979, 32). Tuna processors were purchasing tuna vessels in order to ensure the availability of tuna for their canneries. According to Joseph:

> Since 1966 the fishery in the eastern Pacific has changed remarkedly.... The fleet has increased nearly three times ... competition has increased sharply and the open season for yellowfin fishing has decreased from about 10 months to less than three. Developing nations maintain that under the present management system their tuna fisheries cannot develop, and there are strong pressures for increased special allocations. As these allocations are established there is beginning to be a shift of flag vessels from the nations with large fleets to the nations with small fleets. Each nation with vessels fishing in the CYRA is responsible for establishing and enforcing its own tuna regulations based on the recommendations of the IATTC. As the vessels relocate themselves in other countries the number of nations involved in the fishery increases, along with the problems of implementation and enforcement. (1973, 14–15)

When the yellowfin catch was reached in the CYRA, vessels moved to other areas where catch restrictions did not apply. Increased fishing outside the CYRA was accompanied by a trend toward tunaboat reflagging. In 1977 ten U.S. tuna vessels transferred to foreign registry and six others applied to the U.S. Maritime Administration (MARAD) for transfers. MARAD granted the transfers on the condition that the

> vessels adhere to the same protective measures that apply to U.S.-flag vessels to reduce the porpoise mortality in association with fishing for yellow-fin tuna. Seiners had to post bonds to assure that they would live up to the regulations with respect to employment of fine-

> mesh aprons, backing-down the end of the net, and other procedures for the escapement of trapped porpoises. Although vessels that transfer to foreign registry will be required to conform to these U.S. standards, they no longer will be bound by the U.S. quota on porpoise kills. (NMFS 1977, 12)

Joseph and Greenough comment on the difficulties in implementing national tuna-dolphin regulations for an international resource without coordinated cooperation between nations. Since tuna are a migratory fish, they move from waters within national juridical zones into the high seas beyond national zones. The increase in national purse-seine fishing fleets in the ETP and beyond creates regulation problems, as the dolphin mortality problem is common to all vessels who fish in and near the ETP. "If only one nation enforces strict regulations for protection of porpoise, then this could provide at least partial incentive for vessels of that nation to transfer to flags of other nations without such regulations" (Joseph and Greenough 1979, 169). Extensive reflagging to nonregulated fleets could actually increase dolphin kills.

When Ecuador, Mexico, and Costa Rica withdrew from IATTC in 1980, "the overall effectiveness of the quota was reduced so severely that they were abandoned" (USITC 1986, 103). In 1978 the United States entered into new negotiations with Costa Rica and Mexico regarding the allocation of tuna resources. The coastal Latin American countries wanted guaranteed tuna allocations based on the fact that much of the tuna resources existed within their 200-mile EEZ. Costa Rica and Mexico insisted that such allocations for resource adjacent nations (RAN) should total 69 percent of the catch, while the United States wanted only a 45 percent guarantee. The negotiations deadlocked on the allocation issue with no compromise reached. As a result of the impasse regarding allocations, Mexican seizures of U.S. tuna vessels, which occurred occasionally in the past, increased dramatically in 1980 (USITC 1986, 103).

Exclusive Economic Zones

Tuna harvesting in the ETP has experienced long-term structural changes as various countries developed tuna industries that increased the conflict over access to tuna supplies. With the adoption of purse seiners in the late 1950s, the U.S. tuna fleet expanded its operations from Mexico to Chile. Prior to the mid-1970s most Latin American coastal nations claimed either a three-mile or a twelve-mile national fishing zone. These national fishing zones presented few problems to the U.S. fleet, as most of the tuna fisheries were outside these limits. During the 1970s, however, Mexico (1976), Guatemala (1976), Honduras (1980), Costa Rica (1975), and Colombia (1978) extended their exclusive economic zones to 200 miles (USITC 1986, 102).

The United States and Japan initially did not recognize these claims for tuna, which is a pelagic species. Since the tuna harvested by U.S. tuna fishermen were found almost entirely within the EEZs of Latin American countries, the closing of those fisheries "effectively excluded U.S. fishermen from their traditional tuna-harvesting grounds" (USITC 1986, 133). Efforts by the U.S. government and tuna fishermen to negotiate new access agreements with the Latin American countries initially failed. The Latin American countries' development of national tuna fisheries increased competition for tuna resources and consumer markets. As a result, several American tuna vessels exited the U.S. tuna fleet and transferred to the national flags of Mexico, Venezuela, Grand Cayman, Chile, and Korea (USITC 1986, 147). They also began to move their operations to the WTP.

The Pacific Islands region is located in the WTP and includes fourteen independent and self-governing countries and eight British, French, New Zealand, and U.S. territories. The independent countries are members of the region's political grouping, the South Pacific Forum (SPF), and its Solomon Islands–based fisheries development and management arm, the Forum Fisheries Agency (FFA). Between 1977 and 1984 all countries and territories in the region declared EEZs that accounted for about two-thirds of the total Pacific Islands region. The EEZ were developed to garner revenue from fees charged to DWFNs and to provide some protection for the development of national fisheries. DWFNs arrange tuna harvesting contracts with island nations. "Distant-water fishermen benefit from these arrangements by gaining secure and stable access to rich fishing grounds, while island countries derive revenue and, in some cases, fisheries aid" (Doulman 1987, 36).

The U.S. fleets' relocation to the WTP from the ETP initially resulted in the seizure of several American vessels because of EEZ violations. The United States then enacted embargoes against the offending nations.

After a long series of negotiations between the United States and the Pacific island nations, an agreement was concluded that provided access to the EEZ fishing grounds for U.S. vessels (Iversen 1987a, 271).

The Development of the Mexican Tuna Industry

In the mid-1970s Mexico began to build its modern tuna fleet. By October 1981, Mexico had a fleet of about sixty vessels and had placed orders for an additional sixty-one. These vessels were to be built in sixteen different shipyards and were scheduled to be delivered at the end of 1982. Most of the vessels were purse seiners, with twenty-one to be built in Spain, fourteen in Mexico, seven in Italy, four in the United States, two in Norway, and two in Canada. This fleet would then be the second largest tuna fleet in the world; the largest was the American purse-seine fleet. The Mexican Department of Fisheries also began building modern tuna ports and canneries in southern Mexico. Some of the facilities were to be operated by Mexico's state-owned fishing company, Productos Pesqueros Mexicanos (PPM), which had obtained loans from four Mexican government agencies, Saudi Arabia, and Denmark (*Marine Fisheries Review* 1982b). By 1987 the Mexican tuna fleet was the largest in the world (Hudgins 1987, 153; Joseph 1986).

The 1981–1982 Mexican fleet additions were the largest in the history of the ETP tuna fishery. NMFS officials stated that the Mexican buildup would place even greater stress on the already overfished tuna stocks in the ETP; yellowfin yields and average fish size had been decreasing for several years. Although the Mexican government also expanded old canneries and built new ones to unload, process, and store the increased tuna catch, "it is unlikely that these facilities, however, will be capable of handling the catch of all the new vessels in the near future" (*Marine Fisheries Review* 1982b, 29). According to NMFS officials:

> The problem will be especially acute if Mexico and the United States do not resolve their dispute over tuna management policies so that Mexico can resume exporting frozen tuna to the United States. Plans for the development of the Mexican tuna industry were based on exporting to the United States, one of the world's largest tuna markets. A continuation of the U.S. embargo on Mexican tuna for a

> prolonged period will significantly affect Mexico's economic calculations. (*Marine Fisheries Review* 1982b, 29)

In January 1980 President Jorgé Lopez-Portillo issued a decree requiring license fees for tuna fishing in Mexico's EEZ. During 1980 Mexico seized fifteen American tuna vessels, which were charged and fined for illegally fishing for tuna in Mexico's 200-mile EEZ (Mexico established its EEZ in 1976). Under Section 205 of the Magnuson Fishery Conservation and Management Act of 1979, the United States embargoed all Mexican tuna imports from U.S. markets for two years; this was standard policy in response to vessel seizure. At the time, the United States did not recognize coastal state jurisdiction over highly migratory species such as tuna beyond twelve miles (Hudgins 1987, 156; *Marine Fisheries Review* 1982a, 29; USITC 1986, 101, 104).

The last Mexican seizure of an American vessel occurred in 1983. In 1985 the U.S. State Department began exploring the possibility of ending the embargo. After several months of negotiations between the American and Mexican governments, in August 1986 the embargo was terminated. Because of the U.S. tuna industry's concern regarding the possible impact of Mexico's entry into the U.S. tuna market, the discussions focused on the development of an orderly marketing agreement to "alleviate the short-run effects of terminating the embargo" (USITC 1986, 104). According to Hudgins:

> In Mexico City the "tuna war" with the United States was seen as an issue of national integrity, while enforcement of the U.S. embargo was primarily motivated by domestic protectionism. In April of 1986 Mexico proposed voluntary annual export restraints of 20,000 tonnes to the U.S. raw tuna market. Government sources in both countries believe that this proposal provided the impetus for resolution of the embargo in August 1986. (1987, 156)

The embargo was not the only factor that constrained Mexico's attempt to develop a modern tuna industry. Domestic economic conditions in Mexico, such as high inflation and interest rates, currency devaluations and depreciation, and increasing external debt, also curtailed planned expansion of fishing vessels, canneries, and support facilities. Abundant world tuna supplies (especially in 1983) also depressed world tuna prices, which affected the profitability of the Mexican tuna industry. The advent of El Niño in 1983 reduced tuna catches in 1984, which contributed to a decrease in the number of active Mexican tuna vessels (USITC 1986, 105).

The growing debt on Mexico's new fleet further complicated the industry's development. The ownership of the Mexican fleet is divided into three groups: private companies, state companies, and cooperatives. The private companies usually consist of investment groups or individual investors and also often include considerable investment in joint-venture agreements. The state companies are part of the government-controlled PPM, which is administered by the Subsecretario de Pesca (Subsecretariat of Fisheries [SEPESCA]) (Hudgins 1987).

PPM was formed in 1971 and is composed of several subsidiary companies that operate fishing craft and processing plants throughout Mexico. It is involved in the harvesting, production, and marketing of a wide range of fishery products such as tuna, shrimp, and sardines. In 1980 PPM entered the Mexican tuna harvesting sector with the purchase of two purse seiners. The cooperatives that began operating in 1971 are composed mainly of individual fishermen who collectively own and operate the tuna vessels. In 1984 there were twelve state companies, twenty-four cooperatives, and thirty-three private companies in the Mexican tuna industry (USITC 1986, 107).

The Mexican government banking system is also involved in the tuna industry. Banco Nacional Pesquero y Porturio S.N.C. (National Fishery and Ports Development Bank [BANPESCA]) is primarily responsible for providing relatively low-interest loans and loan guarantees for the construction and improvement of tuna fishing vessels and processing plants. In the mid-1980s BANPESCA was also involved in restructuring tuna vessel debt and providing vessel operating loans. By 1986 BANPESCA owned ten purse seiners as a result of foreclosures or delivery by foreign shipyards (USITC 1986, 123). Mexican government expenditures in support of the tuna industry totaled 140.214 billion pesos ($835 million) in 1984 (USITC 1986, 121). "National industrial policy is developed by SEPESCA, with BANPESCA providing funds to the projects identified by SEPESCA. BANPESCA was specifically established in 1979 to provide credit to the country's fishing industry and to facilitate the financing of new fleets and processing plants" (Hudgins 1987, 154).

A major factor that impacted the expansion of Mexico's tuna industry was the Mexican government's decision to withdraw from the world currency market from February to June of 1982. This action effectively abandoned support of the peso value against the U.S. dollar, resulting in a 44 percent decline in peso value in February alone. The Mexican government withdrew from the world currency market again in August 1982; it also nationalized private banks and established a system of exchange controls in September 1982 to try to stem capital flight and to ration hard currency, especially U.S. dollars. Since many of the new purse seiners were ordered on contracts that specified payment in dollars, the devaluation of the peso increased the real cost of the tuna vessels as well as much of the equipment to operate and maintain tuna vessels and processing plants, all of which were manufactured in the United States and had to be imported (USITC 1986, 109).

The debt for all but one of the new tuna vessels scheduled for delivery in the early 1980s was in U.S. dollars. The vessels were ordered before the currency devaluation with delivery dates after the devaluation. By 1982 the effects of the currency devaluation combined with falling oil prices and revenues and a large external debt ($80 + billion) "drove the Mexican economy into a fiscal crisis and severe recession" (Hudgins 1987, 156). Several tuna vessels went out of service in 1982–1983. The peso debt burden on the new purse seiners increased by over 700 percent between 1980 and 1984. During the same time period, the world tuna price declined by over 30 percent (Hudgins 1987, 156).

In 1986 BANPESCA worked out an agreement with vessel owners to assume the external debt with the foreign shipyards. BANPESCA refinanced the vessels with the Mexican owners at an average of 175.5 pesos to the dollar; the exchange rate at the time of the purchases in 1980 was 22–25 pesos to the U.S. dollar but had climbed to over 300 pesos by 1984. Refinancing required the tuna vessel to meet production quotas to qualify for continuing participation in the program. Few vessels were able to meet the quotas. "Repayment of the vessel debt is a continuing source of friction between government and industry" (Hudgins 1987, 157).

The Mexican government's 1976 establishment of the EEZ, the 1980 declaration of fee licenses for DWFN to fish the Mexican EEZ, and the activities of SEPESCA and BANPESCA "signaled the Mexican private sector that the government would support investment in fisheries" (Hudgins 1987, 162). Investors responded with new purse-seine orders. Enforcement of the EEZ led to U.S. boat seizures and the ensuing embargo that shut off Mexico from the world's largest tuna market. Currency devaluation led to rapid increases in vessel debt load, which the government refinanced. "It is still debatable why the embargo lasted as long as it did, but clearly the Mexican government policy was in conflict with industry development goals (Hudgins 1987, 166).

New Technology, Industry Expansion, and Challenges

The initial wave of expansion of the tuna industry was based on the implementation of purse-seine nets and porpoise fishing. The increases in productivity allowed the growth of the industry in an economically strong region such as the United States. High wages, expanding employment, and low-cost products were all characteristics of an industry centered on a high Fordist production system. Although the expansion of tuna fishing generated high levels of accumulation, it resulted in the emergence of resistance at various levels. Domestically, this production system was challenged by environmental groups, who reacted to the high levels of dolphin mortality associated with the fishing technology. They demanded changes and, more importantly, the cessation of dolphin mortality. This resistance created a crisis of legitimation, as industry actions were increasingly viewed as unacceptable by segments of the U.S. population. The U.S. nation-state mediated the interests of the tuna industry, centering around capital accumulation, and the interests of the environmentalists, who challenged the legitimacy of the fishing technique, with the passage of the Marine Mammal Protection Act of 1972. Internationally, the high profit levels associated with the tuna industry's expansion attracted non-American producers, which translated into the intervention of foreign nation-states in the processes of promoting the economic expansion of national tuna fishing industries and protecting them from external competition. The expansion of developing countries' exclusive economic zones is a case in point. Although at the beginning of this period American-based tuna companies had enjoyed free access to the lucrative ETP tuna fishery off the coasts of several Latin American countries, by the end of the period the ability of the U.S. state to protect the interests of U.S. companies was successfully challenged.

5
THE RESTRUCTURING OF THE GLOBAL TUNA INDUSTRY

> The restructuring of world tuna markets essentially began when multinational corporations moved out of their own countries to conduct tuna business in foreign countries. Before 1980 tuna production and consumption were usually integrated in the same country. That is, the U.S. tuna fleet caught the tuna that was ultimately canned by U.S. processors and consumed by American consumers. The Japanese fleet likewise caught most of the tuna consumed by Japanese consumers. Since 1980, however, tuna exchanges have taken place in a more sophisticated international environment. Today, the tuna consumed by U.S. consumers may have been harvested in the western Pacific, transshipped to Thailand, canned in a plant leased to a multinational corporation, labeled, and distributed by yet another multinational. Tuna harvested by the Japanese fleet may be shipped or traded worldwide by "maguro shosha," specialist tuna-trading companies. (Hudgins and Fernandez 1987, 289)

As indicated in Chapter 2, the period covering the mid-1970s through the early 1980s simultaneously contains the most pronounced crisis of Fordism and the consolidation of emergent global post-Fordist characteristics. The tuna fishing industry was not immune to these changes. The inability of nation-states to legitimize accumulation strategies led to the globalization of tuna operations and the redistribution of power among relevant players. In this chapter we focus on the structural transformation in the industry from its Fordist character to its contemporary post-Fordist organizational form. We articulate this transition through an illustration of the effects of the global economic downturn of the early 1980s, the rise of Third World low-cost tuna production, the decoupling of U.S.-based firms from U.S.-based fleets, the shift to Asia of ownership and localities of operation of prominent tuna TNCs, and, finally, the impact of the restructuring process on U.S. production plants and labor.

The Economic Downturn of 1981 and the Restructuring of the Tuna Market

According to King (1987), the early 1980s slump in world tuna prices was caused by a combination of both supply and demand changes. The supply-side effects included (1) the emergence of productive purse-seine tuna fisheries in the central/western Pacific, (2) overcapacity in the global tuna fleet, (3) a strong U.S. dollar, and (4) a few years of very large skipjack catches. The demand-side shifts were caused by (1) recessionary economic conditions, (2) low relative beef and poultry prices, (3) and the pricing of canned tuna products beyond the apparent "psychological threshold" of $1 per can (for chunk light) (King 1987, 68).

During 1982 there was a substantial decline in U.S. canned tuna sales. Poor consumer demand from 1981 to 1986 produced a downturn in the international tuna market, and fishermen and processors experienced increased difficulties in operating profitably. This situation "forced production adjustments on tuna fishing fleets" throughout the world (Doulman 1987, 9). According to King (1987, 65), in the early 1980s overproduction contributed to a saturation of the market, which resulted in a 7 percent decline in the international price of tuna during the 1981–1983 period. The international economic conjuncture was characterized, he continues, by increasing fuel, labor, and finance costs, which contributed to the elimination of many U.S. tuna producers from the market. Simultaneously, taking advantage of a strong dollar, non-U.S. producers were able "to withstand the cost-price squeeze better than U.S. producers, and as a result they penetrated the U.S. raw/frozen and canned tuna markets. This, in turn, contributed to the geographic redistribution of tuna fishing and helped start the restructuring of the world tuna industry that is currently under way" (1987, 65). The causes of the shifts in production arrangements include the rise of EEZs, exchange rate fluctuations, and the world recession that "caused corporations to seek lower wages and production costs in foreign countries" (Hudgins and Fernandez 1987, 290).

Corporations that had vertically integrated sold their tuna vessels and imported lower cost foreign-caught or -processed tuna. Although global economic restructuring has produced new economic arrangements and new industries have emerged, "The same corporations dominate transactions in harvesting, transshipping, processing (canning/drying/freezing), distribution, and trading. Almost 70 percent of all tuna and tuna products today is caught, processed, traded, or distributed by 10 multinational corporations. These corporations operate on every continent and at every level of the market" (Hudgins and Fernandez 1987, 290).

Table 5.1 lists activities by selected location for the ten tuna TNCs that have dominated world tuna markets at every level of production. These corporations' strategies are generally similar. For example, they are divesting themselves of vessel equity of any sort, including financing. Increasing competition forced many vessels to lose their contracts as a result of the processors' increased use of the world spot tuna market to buy tuna at lower prices. "This divestiture will inevitably lead to further declines in some established fleets while opening the door to the entry of others" (Hudgins and Fernandez 1987, 290).

Although transnational corporations cut their integration with their national fleets, they have expanded operations into transshipping facilities. Increased catches from the WTP are collected for transport to Puerto Rico, American Samoa, the Philippines, or Thailand. By 1987, the principal transshipping point for tuna imported to the United States was the Seychelles, the base of the French and Spanish purse-seine fleets operating in the Indian Ocean. South Korea and the Ivory Coast were the second and third most utilized transshipping points (Parks et al. 1990, 19).

Table 5.2 shows past and current tuna industry activities for TNCs by market involvement segments. TNCs are able to use mobility and international linkages to foster their interests. For example, Safcol is a TNC with six facilities in Australia and seven in Indonesia and the Philippines. At the same time, Star Kist maintains outsourcing contracts from canneries in Thailand, while Mitsubishi leases production facilities in Puerto Rico. Both leasing and outsourcing provide TNCs with the necessary "flexibility to respond to changes in markets, resource availability, or tax/tariff considerations. All but one processing plant has closed on the U.S.

Table 5.1. Major Corporate Entities in International Tuna Markets by Past and Current Activity and Location

	Bumble Bee	C. Itoh	H. J. Heinz (Australia)	John West	Marubeni	Mitsubishi	Mitsui	Safcol	Star Kist	Van Camp
American Samoa									123	123
Australia		1234	1234	4	12			1234		
Canada									34	
Ecuador	123									
Fiji		14								
France		2							2	
Ghana									2	2
Guam									2	
Indonesia		12							2	
Ivory Coast									2	2
Japan	45	12345			12345	12345	12345			
Mauritius						23				
Philippines		1						3		
Puerto Rico	123					3	3		123	123
Réunion Island									2	
Senegal									2	
Thailand								23		
Trinidad										2
United Kingdom				4						
United States	134					4	4	5	134	134

Key: 1 = harvesting; 2 = transshipping/cold storage; 3 = processing; 4 = distribution; and 5 = trading.
Source: Hudgins and Fernandez 1987, 292.

Table 5.2. Major Corporate Entities in International Tuna Markets by Past or Current Level of Market Involvement, 1987

	Harvesting	Transshipment	Processing	Distribution	Trading
Bumble Bee	x	x	x	x	
Mitsubishi	x	x	x	x	x
Marubeni	x	x	x	x	x
Mitsui	x	x	x	x	x
Safcol	x	x	x	x	x
Star Kist	x	x	x	x	
H. J. Heinz (Australia)	x	x	x	x	
Van Camp	x	x	x	x	
C. Itoh	x	x	x	x	x
John West			x	x	

Sources: Ashenden and Kitson 1987; Comitini 1987; Crough 1987; Iversen 1987b.

mainland, with production predominantly shifted to American Samoa and Puerto Rico" (Hudgins and Fernandez 1987, 29).

When planning their tuna activities, TNCs respond to policies or regulations that affect their access to harvesting, processing, and marketing resources or enhance or detract from their profit positions in the market. For example, the fact that the Australian government manages southern bluefin tuna catches with quotas has "influenced processing corporations to shift out of vessel operations and toward financing" (Hudgins and Fernandez 1987, 294). The favorable position that the Thai Board of Trade gives export-oriented firms influences Safcol's operations in Thailand. Star Kist and Van Camp's large operations in Puerto Rico and American Samoa take advantage of favorable tax and tariff regulations as well as special legislation that allows foreign vessels to land at these ports. Furthermore, tuna landed at American Samoa is granted tariff-free entry into the U.S. market (Hudgins and Fernandez 1987, 294). Finally, in their attempt to increase profit margins, the Japanese specialist tuna trading companies are decreasing their joint-venture investments in harvesting and production and focusing on marketing and distribution. According to Hudgins and Fernandez:

> The corporations overall seem to be affected more by political and economic risk factors than by ownership barriers, for example, requiring a certain percentage of ownership by nationals. If all else

> is equal, the corporations appear more willing to locate in known legal situations such as American Samoa and Puerto Rico. Australian firms have longer-term relations with Asian markets and therefore locate there. Exceptions to this pattern exist, but are still experimental, like the Manta, Ecuador, plant. (1987, 294)

The following is a synthesis of the corporate descriptions of the major tuna TNCs presented by Hudgins and Fernandez (1987). Star Kist Food is a subsidiary of H. J. Heinz Company, with about a one-third share of the U.S. tuna market. In 1986 Star Kist's gross revenues were $829.6 million, 19 percent of Heinz's total earnings. Star Kist operates the world's first- and second-largest tuna processing operations in Puerto Rico and American Samoa, respectively. The Mayaquez, Puerto Rico, plant operates at 100 percent efficiency, processes about 187,000 tons per year, and employs 4,000 people at the minimum wage of $3.35 per hour. The American Samoa plant in Pago Pago operates at 90 percent capacity, processes about 102,000 tons per year, and employs 2,358 people at $2.82 per hour. Star Kist also operates cold-storage facilities in France, Ghana, Ivory Coast, Réunion Island, and Senegal, as well as a processing plant in Douarnenez, France. To avoid financial risk at the harvesting level, Star Kist has reduced its equity in purse seiners from 50 to 20.

Van Camp Seafood is the second-largest U.S. tuna company, with about 16 percent of the market for its Chicken of the Sea label. It was a subsidiary of Ralston Purina until 1985, when it became independent. In 1985 its earnings were $270.8 million. Its Mayaguez, Puerto Rico, plant processes about 50,000 tons annually and employs 1,244 people at $4.40 per hour and another 1,226 people at $2.82 per hour. Its plant in American Samoa processes 70,000 tons per year. Van Camp also maintains cold-storage facilities in Ghana, the Ivory Coast, and Trinidad.

Bumble Bee is the third largest tuna corporation in the United States. It maintains a subsidiary, Bumble Bee International, in Tokyo and operates tuna processing plants in Mayaguez, Puerto Rico, and Manta, Ecuador. The Puerto Rico plant operates at 93 percent efficiency, processes about 64,000 tons of tuna annually, and employs 1,300 persons at $3.97 per hour. After a long history as a subsidiary of Castle and Cooke, Bumble Bee management bought the company in 1985 and kept it a private corporation. Gross revenues for tuna operations were $213.5 million in 1984. Bumble Bee owns two purse seiners operating in Ecuador.

H. J. Heinz Company (Australia) was established in 1935 and is an independent subsidiary of its parent, H. J. Heinz, USA. Its 1985 earnings from all of its products was $179 million. This company handles Greenseas tuna labels, 9-Lives pet food, and Epicure/Golden Days health food products. Heinz's attempts to expand in Australia include the purchase of the brand names of Seahaven and Frelish tuna from West Ocean Canning and Heinz and Safcol provided aerial spotting and shore facilities to enable Australian vessels to increase their domestic catches during the 1970s. Heinz accounts for the majority of the Australian tuna market (USITC 1986, 22).

Safcol Holdings is Australia's largest tuna processor and trading group. Safcol has six processing plants in Australia and seven in Thailand. In 1985 Safcol–Australia earned about $83 million and Safcol–Thailand about $63 million. Safcol is a subsidiary of the Southern Farmers Group, Australia's third-largest food processor, and Southern Farmers Group is part of Industry Equity, the fourth-largest company in Australia. Safcol imports all the raw material for its Thai operations. Safcol's recently purchased subsidiary, Interpesco, based in San Diego, California, purchases raw/frozen tuna on the world market and handles the distribution of Thai tuna products in the United States. In response to the yen's appreciation, in mid-1960 Safcol entered the Japanese canned tuna market. Contracts were negotiated directly with supermarkets, which sold the tuna under their house labels. This arrangement bypassed the usual wholesale market and avoided elaborate trading company arrangements and fees.

Safcol closed its Port Lincoln and Melbourne plants in the mid-1980s and now centers its Australian operation in Victoria. For harvesting, Safcol charters a New Zealand vessel. Safcol used aerial spotting to try to remain competitive in the early 1980s but southern bluefin quotas reduced domestic catches and vessel profits. In 1977 Safcol–Philippines and Indonesia entered a joint venture for Philippine tuna canning with Judric Seafoods Company of Hong Kong, a company favored by the Philippine Board of Business. By 1980 the plant operated twenty-four hours per day and processed 200 tons per day, an amount equal to one-half of the country's tuna exports. The success of this venture led to a second venture by Safcol and Judric called Philippine Tuna Canning Corporation, which operated several canneries. The Philippine tuna supplied U.S. institutional markets and represented the largest percentage of any cannery in the world supplying the United States. The deterioration of the international tuna market prices and poor Philippine government planning made the new operations economically vulnerable. The increased competition and worker unrest in the early 1980s worsened the situation to the point that in 1984 the Manila plant was closed, followed by the Zamboanga plant. Safcol lost $4.9 million in the Philippines in 1984. It also lost money on its operations in Indonesia in the early 1980s, where it closed its plant in 1984. Safcol has since concentrated on the Thailand operations.

Soga Shosha are members of the Japanese specialist tuna traders network. The four top companies—Mitsubishi, Marubeni, Mitsui, and C. Itoh—have been involved in world tuna trade since the 1960s and are among the six largest corporations in the world. They view their function in the tuna industry as (1) arranging transportation, (2) financing, (3) providing documentation formalities, (4) ensuring delivery, and (5) inspecting fish. The recent decline in profit margins and increased risk associated with tuna vessel integration and financing has reduced the number of active soga shosha. As a result, the concentration level of these four firms has increased to about 70 percent of the Japanese sashami market (a specially processed tuna product) as these firms have relied on negotiating whole boatload purchases from Japanese distant water vessels and have increasingly acted as marketing agents for Korean and Taiwanese tuna fishing enterprises. All of these firms have cold-storage facilities and agents at the main Japanese landing ports at Yaizu, Japan.

Mitsubishi Corporation has several subsidiaries in Japan (Mitsubishi Shoji) and the United States (Mitsubishi Foods) that manage tuna distribution. This company's main port operations in Yaizu are handled by Toyo Reizo cold-storage company. Mitsubishi ships from Japan to canning operations in Asia (Taiwan and Thailand) about 10,000 to 15,000 tons annually for processing. Mitsubishi's role as a shareholder in the Pacific Ocean fishing endeavors of Kaigai Gyogyo Kabushiki Kaisha (KGKK Overseas Fishing Company) during the period 1958 to 1983 has generated past joint-venture fishing and transshipping arrangements in Malaysia, Mauritius, Papua New Guinea, and Madagascar. Changing political and economic conditions in most of these locations have reduced Mitsubishi's activity to only the Mauritius joint-venture tuna cannery. Tinian Island in the Western Pacific Ocean serves as Mitsubishi's transshipment base for tuna bought from Japanese, Korean, Taiwanese, and U.S. vessels in the Pacific. It also has a cannery in Ponce, Puerto Rico, with a capacity of 27,000 tons annually that employs about 700 people. Total earnings from tuna operations were about $175 million for 1986.

Marubeni Corporation arranges many tuna vessel shipping contracts for Australian supplies of raw tuna through its sashami market agents. Its subsidiary, Marubeni Australia, had earnings of about $88 million for 1985–1986. Its subsidiary, Marukai Suisan, subcontracts cold storage in Yaizu, and its Tokyo subsidiary, Marubeni Reizo, provides cold storage for tuna products distributed to sashami markets in Japan.

C. Itoh and Company has built several purse seiners in the past but now only purchases tuna from fleets operating in the Pacific, usually Korean and Taiwanese vessels. The company then ships this tuna to and from its Japanese cannery and out to Western European markets. C. Itoh's subsidiary, Yaizu Suisan, provides cold storage, processing, and marketing in Japan. Earnings on tuna for 1986 were about $145 million.

Mitsui and Company has two subsidiaries, Tokyo Commercial Company in Yaizu and Towa Suisan in Tokyo, that handle tuna trading in Japan. Its subsidiary Mitsui and Co. USA operates a wholly owned subsidiary, Ocean Packing Corporation, headquartered in White Plains, New York. Ocean Packing operates a tuna processing and canning facility, Neptune Packing, in Mayaquez, Puerto Rico, that employs about 500 people.

In summary, according to Hudgins and Fernandez: "In the past ten years the production and consumption of tuna have been separated by locations. This restructuring has opened some levels of the tuna industry, particularly harvesting and processing, to new entrants. Established corporations have moved away from harvesting toward trading and distribution. The corporate activity is clearly driven by the profit motive" (1987, 299).

RESTRUCTURING OF THE U.S. TUNA INDUSTRY, 1980 TO 1987

Events on the international and national scenes that began in the mid-1970s and accelerated in the early 1980s "have combined to bring about fundamental changes in the world tuna industry" (Iversen 1987a, 282). These changes are (1) a rapid buildup of the U.S. purse-seine fleet to a peak of about 140 vessels in 1975, followed by a rapid decline in the 1980s; (2) a worldwide recession starting in 1981; (3) an increase in foreign tuna purse-seine fleets that increased the world's tuna supply and reduced prices; (4) changes in the location of U.S. fleet operations due to the development of EEZs and MMPA; and (5) the emergence of large-scale tuna canning in foreign countries, most notably Thailand. For example, by 1987 foreign imports of canned tuna, mostly from Thailand, had risen to 35 percent, forcing U.S. processors to move offshore and close all their mainland canneries except for one because of "relatively high domestic labor costs and falling domestic production" (Iversen 1987a, 282).

According to the U.S. tuna industry, foreign competition in the form of canned tuna imports was the most significant factor contributing to mainland cannery closings and other problems for the U.S. tuna industry. From 1979 through 1984 the amount of canned tuna imported into the United States more than tripled, "and all of the major U.S. tuna companies were putting their nationally advertised labels on imported canned products" (National Research Council 1992, 31). This increase was almost entirely in the form of tuna canned in water, which American consumers prefer and which is subject to a much lower tariff rate than tuna canned in oil. The problem of foreign tuna imports was considered severe enough by the industry that it petitioned the U.S. International Trade Commission (USITC) for tariff relief from canned imports. Although USITC concurred that the U.S. tuna industry was facing a difficult restructuring and that the increases in canned imports were a contributing factor to the difficult economic times, the "majority of Commissioners found that overinvestment in boats, plants, and inventories during a period of exceptionally high interest rates was just as important, if not more so, in bringing about the industry's present economic condition" (USITC 1984). The industry subsequently attempted "to have the U.S. Congress enact protective tariff legislation" but this also failed (Herrick and Koplin 1984b, 29).

U.S. tuna processors obtain their product from two primary sources: landings in American ports from domestic vessels and imports of frozen tuna. The U.S. share of raw tuna imports has fallen from 49 percent in 1979 to 30 percent in 1984. By 1985 most of the U.S. tuna catch was from the WTP (52 percent), and the ETP produced only about one-third of the catch it had during its peak in 1976 (Iversen 1987a). The rise in interest rates associated with the 1981 recession negatively impacted the U.S. purse-seine fleet. From a peak of 140 vessels in 1979, by 1986 there were only 72 active vessels, a 51 percent reduction. Over the same period competition from Japanese, Mexican, and other nations' fleets increased in both the ETP and WTP. In fact, prior to the 1960s, American-based companies dominated the ETP fishery. Since then, fleets based abroad have significantly increased their share of the ETP harvest. "The proportion of the catch taken by U.S. vessels decreased from 90% in 1960 to 32% in 1988 to 11% in 1991, and that taken by Latin American countries bordering the Pacific Ocean increased from 10% to 47% in 1988 and 57% by the end of 1991" (National Research Council 1992, 4).

Increased harvesting led to depressed prices on international markets. Frozen raw tuna was available on international markets cheaper than U.S. producers could buy it from the U.S. fleet or harvest it with their company-owned vessels. "As a result, many seiners today are fishing on an 'open ticket,' have no guaranteed market for their fish, and must negotiate prices with the processors on a per-trip basis, often when they return to port" (Iversen 1987a, 283–84). From 1980 to 1984, domestic tuna landings at U.S. ports dropped 47 percent and the volume of domestically produced canned tuna fell 56 percent. "U.S. processors were faced either with continuing their processing activities both offshore and in California [where they operated expensive canneries] or with consolidating all of their processing offshore where costs were significantly lower" (Iversen 1987a, 284).

Another factor contributing to the canneries' moves offshore was the changing relationship between the processors and the U.S. tuna fleet that had traditionally supplied the mainland canneries. During the 1970s processors vertically integrated to try to guarantee consistent supplies for their California canneries to meet the growing domestic demand. Most purse seiners were either fully owned or partially owned by the processors. The 1981 recession combined with low-cost tuna on the international market prompted the processors to sell their interest in tuna vessels and rely on short-term contracts. By 1983 most U.S.-based purse seiners were operating independently of the processors (Iversen 1987a, 284).

U.S. tuna processors took drastic actions to ensure their survival in the international tuna market, as aggressive foreign competition and the buildup of modern national tuna fleets have threatened their survival. The rapid growth in demand and imports to the United States of tuna not canned in oil has been accompanied by a rapid decline in the canning capacity on the U.S. mainland. At the end of 1982 there were twenty canneries operating in the United States: twelve on the mainland, five in Puerto Rico, two in American Samoa, and one in Hawaii (see Table 5.3).

Table 5.3. U.S. Tuna Canneries, by Plant Location, 1979–1985

	1979	1980	1981	1982	1983	1985
Continental United States	14	12	12	12	3	1
Hawaii	1	1	1	1	1	0
Puerto Rico	5	5	5	5	5	5
American Samoa	2	2	2	2	2	2

Source: National Marine Fisheries Service as cited in USITC 1986, 172.

These numbers reflect a decrease in the total number of plants operating in the United States since 1977 but not a decline in canning capacity. During the 1979–1983 period, eleven tuna processing plants located in the continental United States closed down (USITC 1986, 24). By 1986 there was only one plant left operating on the mainland, Pan Pacific in Terminal Island, California. The plant in Hawaii had also closed, but the plants in Puerto Rico and American Samoa were still active. The mainland plants closed because of high American labor costs, a strong U.S. dollar, rising foreign imports, and the economic recession of 1981 (Floyd 1987, 86). About 2,400 mainland tuna processing jobs were lost when Van Camp closed its San Diego, California, plant and Star Kist closed its Terminal Island plant (Herrick and Koplin 1986, 28).

As canning capacities declined on the mainland, offshore operations in Puerto Rico and American Samoa, where labor costs are lower and tax incentives are generous, increased. In 1985 Star Kist's plant in Puerto Rico added a third shift and Van Camp's plant there reopened after renovation. Although the Bumble Bee plant in Puerto Rico closed due to the financial problems of its parent, Castle and Cooke, it reopened after its managers bought it. Canning operations in American Samoa expanded even more rapidly than Puerto Rico. Both Star Kist and Van Camp renovated their Pago Pago plants and are increasing production. By 1987, 63 percent of domestic tuna deliveries to U.S. canneries were to American Samoa and California and 37 percent were to Puerto Rico (Parks et al. 1990, 19).[1] According to industry sources,

> The shift in production to American Samoa is due to its proximity to the western Pacific tuna fishery, reasonable wage costs, and tax incentives offered by the American Samoan government for offshore investment. American Samoa's exemption from the Nicholson

Table 5.4. Market Shares of U.S. Tuna Processors, September 1986 (in percent)

Brand	Processor	Market Share
Star Kist	Star Kist	36.0
Chicken of the Sea	Van Camp	20.0
Bumble Bee	Bumble Bee	15.5
3-Diamonds	Mitsubishi	3.7
Geisha	Mitsui	2.8
Empress	Mitsui	1.4
Private Labels		17.0
Other		3.6

Source: Selling Areas Marketing, Inc. (1986).

> Act and its duty free access to the United States under provisions of Headnote (3)a of the U.S. Tariff Schedules are other advantages that make American Samoa an attractive site for tuna canning operations. (Floyd 1987, 87)

Despite the advantages provided by Puerto Rican and American Samoan access to the U.S. market, imports of canned tuna packed in water increased almost fourfold from 1979 to 1984. Thailand replaced Japan as the largest importer to the U.S. tuna market, the Philippines is second, and Taiwan is fourth (Floyd 1987, 87).

In 1987 there were four principal U.S. tuna processors: H. J. Heinz Company's Star Kist, Ralston Purina's Van Camp, Bumble Bee Seafoods, and California Home Brands' Pan Pacific Fisheries. Two Japanese corporations also operated in Puerto Rico: Mitsubishi Foods, a subsidiary of Mitsubishi International Corporation, operated the Caribe Tuna cannery, and Mitsui and Company's subsidiary Ocean Packing Corporation operated Neptune Packing. Market shares of U.S. tuna trade for these firms is reported in Table 5.4. Processing plant locations in 1985 are reported in Table 5.5. A partial review of these firms is provided in the previous section on global restructuring. Other pertinent points regarding the specifically U.S. operations of these firms are discussed below (from Iversen 1987a, 271–74).

Star Kist was founded in 1917 and bought in 1963 by H. J. Heinz Company, a Pennsylvania-based food processing MNC. Through its several wholly owned subsidiaries, it is the largest U.S. tuna processor and producer of canned tuna and tuna-related products such as pet food (9 Lives), with over one-third of the U.S. market. Tuna and related products

Table 5.5. U.S. Canned Tuna Processors, Location by Firms and Processing Plants, 1985

Firm	U.S. Processing Plants
Bumble Bee Seafood, Inc. San Diego, California	Mayaquez, Puerto Rico
CHB Foods Inc.-Pan Pacific Fisheries Terminal Island, California	Terminal Island, California
Mitsubishi Foods Inc. (Caribe Tuna) Delmar, California (a subsidiary of Mitsubishi Corp., Tokyo, Japan)	Ponce, Puerto Rico
Neptune Packing Corp. White Plains, New York (a subsidiary of Mitsui USA, New York)	Mayaquez, Puerto Rico
Star-Kist Foods, Inc. California (a subsidiary of H. J. Heinz Company, Pittsburgh, Pennsylvania)	Mayaquez, Puerto Rico; Pago Pago, American Samoa
Van Camp Seafood Division (a subsidiary of Ralston Purina Company, St. Louis, Missouri)	Pago Pago, American Samoa; Ponce, Puerto Rico

Source: Compiled from information submitted in response to questionnaires of the U.S. International Trade Commission as cited in USITC 1986, 171.

make up the largest share of Heinz's total product sales, about 20 percent. Star Kist has consistently been one of Heinz's most profitable subsidiaries, contributing an average of 23 percent to H. J. Heinz's international earnings during 1984–1986. The Star Kist canneries in Puerto Rico and American Samoa are the largest in the world, and Star Kist also operates a cannery in Canada. In 1984, "in response to continued high costs and the Government's failure to provide relief from low-priced canned tuna imports," Star Kist closed its Terminal Island, California, tuna processing plant (Heinz 1985, 17).

Van Camp Seafood Company was bought in 1963 by the Ralston Purina Company of St. Louis, Missouri, and is a wholly owned subsidiary. It is the second largest U.S. tuna and related products company, with about 20 percent of the market. Van Camp's seafood products accounted for an average 6 percent of Ralston Purina's sales from 1981–1986. Van Camp's operating income was low over the 1982–1986 period, including a $16-million loss on tuna vessels. Van Camp has canneries in Puerto Rico and American Samoa and distributes tuna under its Chicken of the Sea label. It has a transshipping station in Ghana for Atlantic tuna catches and transships through Guam and Tinian for Pacific catches. "In July 1984, Van Camp closed its San Diego tuna-processing plant, said to be the most modern in the world, and transferred its capacity to the company's offshore locations. At the end of fiscal year 1985, Van Camp had equity interests in 15 tuna purse seiners, 8 of which were wholly owned and 1 was leased" (USITC 1986, 23).

Bumble Bee was recently bought from its parent, Castle and Cooke, by its managers. It is the third-largest U.S. tuna corporation, with about a 16 percent market share. When Bumble Bee was a subsidiary of Castle and Cooke, its revenues accounted for an average of 14 percent of the parent company's total revenues from 1982 to 1984. Bumble Bee had negative profits from 1982 to 1984, $12 million, $7 million, and $1 million, respectively. The company has tuna canneries in Puerto Rico and Ecuador. "Bumble Bee began processing fish in 1899 and has been in operations as a tuna processor since 1937, when it started canning albacore in Oregon" (USITC 1986, 23). "In 1982, the firm closed its San Diego tuna plant, a facility purchased in 1979. In 1982, Bumble Bee sold all of its 12 vessels and currently relies exclusively on short-term contracts with U.S. vessels and foreign supplies of frozen tuna to operate its canneries" (USITC 1986, 23).

Pan Pacific Fisheries, a division of California Home Brands (CHB), is the fourth-largest U.S. tuna canner. It sells tuna under its own labels—C.H.B., Top Wave, and Lucky Strike—as well as other labels. Pan Pacific Fisheries accounted for about 30 percent of its parent company's total revenues in 1984. Operating profits for Pan Pacific fluctuated from a high of $7 million in 1983 to a low of minus $3 million in 1984. It operates one tuna processing plant in California, which employs 500 workers, and has no overseas operations. In 1986 Pan Pacific paid its workers union wages of $6.63 per hour. In late 1984, those workers had taken a $1.00 pay cut, about 15 percent of their wage, to be restored over three years (*Seafood Business Report* 1985, 10). In an attempt to attract buyers to a new brand of its tuna, Pan Pacific included this statement in an advertising campaign: "Low-cost labor in foreign countries led all but

Table 5.6. Number and Capacity of U.S. Baitboats and Purse Seiners in Selected Years, 1955–1985 (capacity in short tons)

	Baitboats[ab]		Purse Seiners[b]		Total Fleet	
Year	Quantity	Capacity	Quantity	Capacity	Quantity	Capacity
1955	183	41,729	66	8,250	249	49,979
1960	80	15,691	111	24,971	191	40,662
1965	50	4,279	116	39,611	166	43,890
1970	45	3,852	118	55,823	163	59,675
1975	55	5,483	140	115,342	195	120,825
1980	25	2,186	117	107,734	142	109,535
1985	9	696	90	97,131	99	97,827

[a] Baitboats are limited to those engaged in the Eastern Tropical Pacific.
[b] Consists of baitboats with a minimum capacity of 50 tons and purse seiners with a minimum capacity of 101 tons.

Source: Official statistics of the Inter-American Tropical Tuna Commission; submitted to the U.S. International Trade Commission by the American Tunaboat Association as cited in USITC 1986, 157.

one tuna company to abandon [continental U.S.] facilities and workers" (*Seafood Business Report* 1985, 10).

Mitsubishi Foods is headquartered in California and operates one cannery in Puerto Rico that employs 700 people. It sells its 3-Diamonds brand in the eastern and midwestern regions of the United States. Canned tuna makes up about 50 percent of Mitsubishi food sales; the remainder is in other canned fruits and vegetables. It does not own or operate any fishing vessels. Mitsui and Company operates a cannery in Puerto Rico employing 500 people under its Ocean Packing Corporation.

In addition to the relocation of processing, between 1979 and 1985, the number of U.S. flag purse seiners declined from 124 to 90, and fleet capacity declined from 114,000 short tons to 97,000 short tons. During the same period, the number of baitboats declined from twenty-eight to nine vessels and the number of trollers that spent at least some of their time in tuna fisheries declined from 660 to 108 (USITC 1986, 20–21)(see Table 5.6). Although the number of vessels declined drastically, the average size of the typical purse seiner rose from 125 tons in 1955 to 1,079 tons in 1979. A total of fifty-seven tuna vessels left the U.S. fleet from 1981 to 1985. Of those vessels, thirty-four transferred to foreign flags to fish for tuna, eight transferred to other non-tuna U.S. fisheries, and fifteen were lost at sea (USITC 1986, 8)(see Table 5.7).

Although the ETP had for a long time been the principal fishing grounds for the U.S. fleet, climatic changes such as the occurrence of El Niño combined with CYRA quotas, U.S. regulations such as MMPA, and EEZ enforcement resulted in a decline in ETP participation and a move to the WTP. "The El Niño influence emphasizes the competitive advantage of a large-scale, flexible fleet of distant-water vessels" (USITC 1986, 132). From a high of 125 U.S. purse seiners fishing the ETP in 1979, by 1985 the number had declined to only 49 vessels, a drop of 61 percent (USITC 1986, 9). As the canneries in Hawaii and California closed, the costs of fishing in the ETP increased. Vessels fishing in the ETP were forced to sell their catches in Puerto Rico and American Samoa, which entailed increased transportation costs. Although most of the U.S. purse seiners are new, large superseiners, a "significant portion of the U.S. fleet still consists of older, small purse seiners located in the eastern tropical Pacific" (USITC 1986, 131). "The resulting increased transportation costs, during a time of rising harvesting costs and declining prices, contributed to the decisions of many U.S. vessel operators to either move their operations to the western Pacific or cease active participation in the U.S. tuna fishery altogether" (USITC 1986, 11).

The most important barrier to entry into tuna harvesting is the cost of the purse seiner. Purse seiners built in the late 1970s and 1980s cost between $5 million and $10 million, an amount beyond the abilities of the average investor. This fact necessitated a co-owner for the vessel,

Table 5.7. U.S. Tuna Purse-Seine Fleet: Fleet Size, Additions During Year, Transfers to Foreign Flags, and Idle Vessels, 1978–1986 (capacity in short tons)

	Fleet size on January 1		Additions During Year		Transfer to Foreign Flag		Idle vessels on Dec. 31	
	No.	Capacity	No.	Capacity	No.	Capacity	No.	Capacity
1978	124	110,665	4	4,800	4	4,061		
1979	122	109,109	5	6,600	1	550		
1980	124	113,589	5	6,000	9	10,500	2	1,292
1981	117	107,349	13	14,750	1	550	11	8,587
1982	128	121,194	11	13,250	8	7,690	19	15,808
1983	125	124,173	6	7,750	5	5,081	34	28,933
1984	124	126,164			11	9,374	23	21,010
1985	107	110,985			9	8,585	21	22,422
1986	90	97,131						

Source: Data submitted by the American Tunaboat Association as cited in USITC 1986, 158.

often a processing firm that had sufficient capital to finance the investment. Since the purse seiner has been built specifically for tuna fishing, retrofitting costs are high. "The nontuna options for a large tuna purse seiner are few" (USITC 1986, 14). The expanding market for purse seiners in foreign countries provided an outlet for excess American ETP vessels. Although occasional flag transfers have been a normal occurrence in the tuna industry, from 1979 to 1985 a total of forty-four U.S. tuna vessels transferred to foreign flags; flag transfers peaked at eleven during 1984 (USITC 1986, 15)(see Table 5.7).

The increase in reflagging that occurred in these years is linked to the increased costs of production in the United States associated with the passage of MMPA and the global economic crisis. A dominant strategy of producers was to nominally register their multimillion-dollar boats to flags of convenience in order to bypass U.S. legislation. "By 1986, only 34 boats of the 103 boat-fleet were registered in the United States and subject to National Marine Fishery Service (NMFS) regulations; by 1991 only 11 US boats fished the ETP" (National Research Council 1992, 4). It is important to note that the personnel employed in these boats are virtually the same as those employed when the boats were U.S. registered. In other words, there has simply been a cosmetic change that eliminated the impact of MMPA to these boats. Simultaneously, because these vessels ceased to be registered in the United States, the number of dolphins killed by U.S. crews decreased. Not surprisingly, the number of dolphins killed in the ETP by "non-U.S. boats" increased dramatically (National Research Council 1992).

As a result of global economic restructuring, relations between the tuna fleet and the processors "have undergone significant changes since the late 1970s" (Iversen 1987a, 274). Before 1980 tuna processors either owned, had controlling financial interests, or had long-term contracts with the U.S. purse seiners that supplied their tuna; this guaranteed them tuna supplies in times of high demand and low supply. The arrangement also provided a secure market for the catches of U.S. tuna fishermen. But as interest rates rose, so did carrying costs for the integrated vessels. "As a result, it became more economical for vertically integrated processors to let independent tuna vessel owners bear the financial risk of operating the vessels" (USITC 1986, 27).

As foreign fleet development increased global catches, world tuna prices fell to the point where sourcing tuna on the spot markets or with short-term contracting was much less costly than formally integrating tuna harvesting. The easy availability of low-cost, imported frozen tuna created downward pressure on the prices negotiated by domestic processors for foreign and U.S. tuna. Presently, prices of tuna from sources are negotiated in very short-term contracts, a sharp contrast to the contracts of several months that were common during the 1970s and 1980s (USITC 1986, 149).

Consequently, U.S. processors sold off their tuna vessels and quit arranging long-term contracts with other vessels, effectively divesting their integration with the U.S. tuna fleet (USITC 1986, 25–26; Iversen 1987a). As of 1986, several vessels still owned by U.S. processors were tied up or for sale (USITC 1986, 13). According to data provided by Iversen (1987a, 274), Bumble Bee, for instance, acquired twelve seiners from the Gann fleet in 1975 for $1 million and resold them in 1982. Additionally, Van Camp in 1986 controlled seventeen seiners, of which it had a 100 percent interest in nine, a 50 percent or less interest in six, and a long-term lease on two. Star Kist had partial ownership of about fifteen or twenty seiners, but at one time it had interest in almost fifty of them. Bumble Bee owns two boats in Ecuador. "In 1984 Pan Pacific owned 11 seiners but recently sold its remaining 6 vessels (2 to a U.S. firm, 3 to Venezuelan interests, and 1 to Ecuador). The end result of this divestment in fleet by the processors was a shift of financial risks associated with vessel operations from the processors and harvesters entirely to the harvesters" (Iversen 1987a, 274). "During this period, many U.S. tuna vessels were sold to nations with lower fuel and labor costs and more advantageous tax climates. This conduct put independent U.S. tuna boats at a competitive disadvantage in terms of harvesting costs during this period when the growing global tuna harvest was holding down raw tuna prices and canners had many new supply sources" (National Research Council 1992, 30).

As a result of deteriorating relationships between the U.S. processors and U.S. fleet, a group of American independent tuna vessel owners sued the three largest processors in 1985 for monopolistic activities; the twenty-four plaintiffs represented fifty-four U.S. tuna purse seiners, a majority of the U.S. fleet (USITC 1986, 26). The vessel owners alleged that "Star-Kist, Ralston Purina (Van Camp), and Castle and Cooke conspired to restrain trade, fix prices, monopolize trade, and destroy the independently owned tuna fleet" (Floyd 1987, 88). The suit was settled out of court in 1986.

Following processor divestiture and vessel movement to the WTP, in 1984 there was a rapid rise in the amount of American harvested tuna exported to foreign canneries. The U.S. fleet exported 29,750 tons of tuna in 1984, compared with 583 tons in 1983. Most of the product was transshipped to Thailand, Japan, and Italy. "The surge in U.S. exports is an indication of the internationalization of tuna trade in general and, more specifically, a possibly increasing reliance of U.S. vessels on foreign markets" (Herrick and Koplin 1986, 28). Indeed, it is entirely possible that the U.S. fleet will be increasingly integrated with foreign tuna TNCs through harvesting contracts.

Tuna processors in American Samoa have an advantage over operators based in Puerto Rico because American Samoa is outside the U.S. customs district, enabling foreign vessels to unload their catches directly into the canneries, while tuna caught by foreign vessels to be landed in Puerto Rico must be transshipped outside Puerto Rico for delivery to the canneries (Iversen 1987a, 275). American Samoa's exemption from the Nicholson Act makes it a very attractive entry point into the U.S. market. According to King,

> the critical question concerning the processing of tuna for the U.S. market is whether U.S. canners operating at their offshore processing sites will be able to compete with Asian processors or whether the move to these offshore sites will be only an interim step in the westward migration of tuna processing operations. U.S. tuna companies continue to increase their purchases of the imported canned product and may eventually stop processing altogether and become US distributors of foreign-produced canned tuna products. Simple cost comparisons suggest that U.S. offshore processing sites cannot compete on a cost basis with Asian processors, so such developments are possible. (1987, 67–68)

Over the period 1979 to 1985, the number of workers in tuna processing fell from 14,668 to 12,887, mainly because of the closings of U.S. mainland plants (USITC 1986, 39–41). In 1987 there were approximately 11,300 persons employed by the four major American MNCs in Puerto Rico, American Samoa, and California. Current wages range from a high of $6.82 per hour for general labor and $8.32 per hour for skilled labor in California (with benefits included, rates are much higher) to a minimum wage of $2.82 per hour in American Samoa. Star Kist's Mayaquez, Puerto Rico, plant pays the minimum wage of $3.35, while Van Camp's plant pays $4.44 and the Bumble Bee plant pays $3.97 (Iversen 1987a; Hudgins and Fernandez 1987). Average hourly wages for all U.S. tuna processing workers increased by 27 percent from 1979 through 1983, from $4.32 to $5.50 per hour; by 1985 the average hourly wage had dropped to $4.82 per hour (USITC 1986, 25). These changes in wages reflect the changes in processing location.

Tuna canning technologies have not changed for decades. The process is labor-intensive, and because of the varying sizes of the fish, mechanization has been limited. The loins (fillets) are removed from the tuna on processing lines, where the light meat used for human consumption is separated from the dark meat used for pet food. The rest of the fish is ground into fish meal and used as a poultry feed supplement (USITC 1986, 18). To try to keep costs down in an arena of increased competition, processors are experimenting with a new two-stage method of tuna processing. Since about 70 percent of the labor costs are associated with getting the loins off the tuna, processors are removing the loins where labor is less costly and transporting them to another area where they are cooked, canned, and marketed. This technique has some drawbacks, especially the short shelf life of loins compared with frozen tuna (Iversen 1987a, 277).

Various governmental legislation and regulations also impact the business activities of the tuna TNCs. The U.S. territories of Puerto Rico, American Samoa, Guam, and the Virgin Islands can ship canned tuna into the U.S. market duty free. Tuna imports are dutied at 6 percent ad valorem on amounts less than 20 percent of the U.S. tuna pack from the previous year; 12.5 percent duty is paid for imports above the 20 percent level. American Samoa is not included in the duty quota equations, which decreases the amount of tuna imported at the lower rate, especially as the California plants closed. Imported tuna canned in oil is subject to a 35 percent ad valorem tariff (Fernandez and Hudgins 1987).

Puerto Rico also provides specific incentives on corporate income taxes, property taxes, and sales taxes. The Industrial Incentive Act of 1978 allows income tax exemptions of up to 90 percent for periods ranging from ten to twenty-five years, depending on their location on the island, to tuna processors and commercial fishing operations that supply Puerto Rican canneries. These exemptions for the five processors are effective through the years 1993–2000, depending on the particular processor. For example, Mitsui's Neptune Packing currently gets an 85.5 percent tax exemption, to be reduced to 35 percent in 1994; Star Kist, Van Camp, and Bumble Bee are in a zone providing the maximum 90 percent tax exemption, although these rates will decrease to 65 percent in 1995 for Van Camp and Bumble Bee and in 1993 for Star Kist; and Mitsubishi's Caribe Tuna's income tax exemptions will be reduced to zero by 1993. Under the law, the processors can renegotiate their levels of tax exemption. Puerto Rico also has a variable toll gate tax that is levied on earnings declared by Puerto Rican–based companies and distributed off the island to parent corporations. If processing companies agree to keep some of the profits in Puerto Rico, the toll gate tax is lower (Iversen 1987a, 278–79).

American Samoa extends tax exemptions for periods of up to ten years for the establishment or enlargement of economic enterprises. Other benefits include tax exemptions for dividends paid by wholly owned subsidiaries of U.S. parent corporations operating in American Samoa and the ability to carry forward business losses for tax purposes for seven years. The territory also has no taxes on sales, property, exports, or value-added items and permits free port storage of up to sixty days for transshipped freight. Both the Samoa Packing Company (Van Camp) and Star Kist Samoa have agreements with the American Samoan government to make investments in increased processing capacity and employment, improve the environmental conditions in Pago Pago harbor, and train American Samoans to be plant managers and purse-seine crewmen. Tuna vessels are also exempt from income taxes. In addition, corporations organized under American Samoan law and whose tuna fleets deliver at least 20 percent of their annual catch to American Samoan canneries are exempt from income taxes (Iversen 1987a, 279).

The U.S. government has been involved in the regulation of the tuna industry since the 1940s. The United States was a founding member of IATTC, which was created in 1949, while the Tuna Convention Act was passed in 1950. Four years later the Fishermen's Protective Act was introduced, and in 1975 the International Commission on the Conservation of Atlantic Tunas and the Atlantic Tuna Conventions Act were introduced. Other major pieces of U.S. legislation that also impact the tuna industry are the Marine Mammal Protection Act of 1972, the Nicholson Act of 1973, which created duty-free status for American

Samoa, and the Fishery Conservation and Management Act of 1976, which extended the American EEZ to 200 miles.

In summary, although U.S. tuna harvesters were the first to adopt the purse-seine technology, reaping substantial profits compared to smaller, less modern, foreign fleets, as other nations developed purse-seine methods, the United States lost its advantage. In order to meet foreign competition, especially from Thailand, and regain profitability, the U.S. tuna processing industry restructured to cut production costs. U.S. tuna TNCs have reduced their integration with the American tuna fleet, eliminated all but one expensive California-based cannery, and relied increasingly on the global spot market for raw tuna resources.

TNCs have consolidated their U.S. operations on Puerto Rico and American Samoa, where U.S. territorial status and tax incentives provide a stable business climate and where they can "take advantage of labor and capital production costs, as well as more liberal worker health, welfare, and environmental regulations" (Fernandez and Hudgins 1987, 145). Although the same general operations are employed to process tuna, "the economics of tuna processing differs substantially from nation to nation, however, because of differences in direct wage rates, the costs of workers' health and safety requirements, environmental regulations, and tax and trade concessions" (National Research Council 1992, 31). During the 1980s, TNCs began shifting operations away from U.S.-controlled territories to "Asian sites to take advantage of even cheaper labor and less costly worker benefits and environmental restrictions" (National Research Council 1992, 31). The American TNCs' disintegration with the U.S. fleet has produced bad feelings between the two groups that resulted in a lawsuit and charges of monopolistic behavior by the vessel owners against the processors.

THE RELOCATION OF THE GLOBAL TUNA FISHING INDUSTRY

Between late 1988 and early 1989, significant portions of the global tuna processing industry were bought by Asian-based firms; Bumble Bee was bought by Unicord, originally based in Thailand, and Van Camp by P. T. Mantrust based in Indonesia. Industry analysts said these transactions reflected the food industry's increasing concerns about costs. According to Nomi Chez, an analyst with Goldman, Sachs and Company in New York, "The food business is consolidating on a worldwide basis and there is a lot of production in the Far East. It's a labor intensive business and labor costs are low there" (Kraul 1989, 2). As a result of these acquisitions, Asian-based TNCs now control over 50 percent of the U.S. tuna market (USITC 1990, 27). The U.S. market represents 31 percent of the total catch, while the Japanese market is about 36 percent (National Research Council 1992, 3). Obviously, the market for tuna products is firmly located in economically affluent societies.

In November 1988 Ralston Purina sold Van Camp (Chicken of the Sea) to Mantrust for about $260 million. By 1990 Van Camp was the second-largest U.S. tuna processor, with 20 percent of the American market. Mantrust, established in 1958, is one of Indonesia's largest food conglomerates, with holdings in agribusiness, distribution, trading, retail industries, and shipbuilding. The firm has three tuna canneries operating in Indonesia and one in American Samoa. In June 1990 Mantrust closed its Ponce, Puerto Rico, processing plant (USITC 1990, 20). Mantrust also has a joint venture factory operation with the U.S. tuna fleet's cooperative association in Bali (*Seafood International* 1988, 35).

Mantrust's Indonesian tuna operations include two fishing fleets under the subsidiary P. T. Nelayan Bahkti, a cannery in east Java under the subsidiary P. T. Blam Bangan Raya, a cannery in Bali under the subsidiary P. T. Bali Raya, and a recently acquired cannery on the island of Biak. Labor costs in Indonesian canneries are based on government minimum wages. In 1990 on the island of Bali, the minimum wage rate was 2,700 rupiah per day, about $1.45; the average monthly wage was approximately 150,000 rupiah, about $80. On Java, the average cannery worker's wage was lower, about 1,500 rupiah per day or 80 cents, which was higher than Java's minimum wage rate (USITC 1990, 84).

Indonesia is one of the world's fastest growing producers and exporters of canned tuna. In 1980 Indonesian tuna exports to the United States were virtually nil, but by 1989 accounted for 7 percent of American canned tuna imports. Although the Indonesian tuna industry presently accounts for a relatively small share of global tuna production and exports, it has a strong potential to become a major global participant. Indonesia has "access and proximity to tuna resources and a huge, relatively low cost labor force. In addition, the recent acquisition of the U.S. tuna processor, Van Camp, by the Indonesian firm P. T. Mantrust gives Indonesia access to improved technology and a large and stable market" (USITC 1990, 82).

When Mantrust bought Van Camp, it bought the technology and market access needed to expand production and trade. Mantrust now dominates Indonesian canned tuna production. It also is one of the world's largest producers of mushrooms and produces baby corn, fruits, dairy products, soft drinks (Pepsi franchise), beer (Anchor brand), beef, and fish products. In addition, the firm holds a manufacturing license for Adidas sports shoes and produces consumer electronic products (USITC 1990, 83).

In August 1989 Pillsbury sold its Bumble Bee subsidiary to Unicord of Thailand for about $269 million. By 1990 Bumble Bee had replaced Van Camp as the second-largest U.S. tuna processor, with 23 percent of the market (USITC 1990, 19). Pillsbury was being restructured under the ownership of Grand Metropolitan, which bought it in January 1989 for $5.7 billion (Kraul 1989). Grand Metropolitan's decision to sell Bumble Bee was motivated by the increasing costs of business in the United States (Handley 1989). In 1985 Castle and Cooke sold Bumble Bee to four Bumble Bee executives for $60 million. In the summer of 1988 the Bumble Bee executives sold the company to Pillsbury for $262 million, after antitrust actions kept Heinz's Star Kist from buying it (Kraul 1989). In 1989, Bumble Bee employed 100 people in its San Diego headquarters, 2,400 at its main cannery in Mayaguez, Puerto Rico, and 250 at another plant in Ecuador (Kraul 1989).

Unicord is a food processing and exporting conglomerate based in Bangkok, Thailand. That country is the world's largest exporter of tuna and was the major supplier of Bumble Bee prior to the 1989 acquisition. Unicord is a subsidiary of a larger corporate group owned by the Thai Konuntaklet family. Unicord relies on imports of raw tuna for its production of canned tuna and "seeks long-term contracts on such procurement to ensure continuous supplies for its production lines" (USITC 1990, 19). In 1988, 48 percent of Unicord's tuna export sales were from the United States, 35 percent from Europe, 8 percent from Japan, and the remainder from the rest of the world (USITC 1990, 19).

Unicord is Thailand's largest exporter of tuna. In 1988, Thai canned tuna exports to the United States were $246 million; total canned tuna exports were about double that amount. In 1988 Unicord canned about 50,000 tons of tuna, with 95 percent of the tuna exported (Handley 1989). "Unicord expects to benefit by locking in Bumble Bee as a U.S. distributer of its tuna" (Kraul 1989, 2). Although Grand Metropolitan sold Bumble Bee to avoid high operating costs in the United States, Unicord's purchase of Bumble Bee was made "to counter stiff U.S. tariffs and quotas on imports of canned tuna and to protect Unicord's stake in the U.S. market" and also to gain vital access to the American market (Handley 1989, 108). Imports of canned tuna packed in oil carry a 35 percent tariff, while imports of tuna packed in brine initially carry a 6 percent duty. Soon after Unicord's purchase of Bumble Bee, it disclosed plans to build a large cannery near San Diego to process Bumble Bee tuna loins (Kraul 1989).

"The importance of securing a good slice of the U.S. market was underlined by the strong turnout for the Bumble Bee auction" (Handley 1989, 108). Bidders for Bumble Bee included Mitsubishi, Mitsui and Company, and Mantrust. By buying Bumble Bee, Unicord can export bulk tuna to plants in California and escape the high duty for canned tuna. Unicord also owns Betagro Feed, poultry, shrimp, and pig farms in Thailand, and American can companies (Handley 1989).

In 1990 Unicord was the world's largest tuna canner. The firm was established in 1978 and by the mid-1980s was Thailand's largest canned tuna exporter. "Bumble Bee was acquired at auction in the first step by Unicord to form a global tuna organization" (Handley 1991a, 48). Unicord's global strategy involves a network of factories in five continents that will give the company easy access both to fish and to its main markets. At the core of the strategy is a new tuna handling process that cuts transport costs and enables Unicord to avoid high import duties in the United States and Europe. Unicord's plan is to ship frozen tuna loins directly to the United States for canning and to discontinue its canning expansions in Thailand. At one point Unicord president Dumri Konuntaklet wanted to close Bumble Bee's high-cost operation in Puerto Rico but changed his mind. He pointed to the loss of U.S. market share by Chicken of the Sea (Mantrust of Indonesia) after closing its Puerto Rico plant and identified Bumble Bee as the means to operate within the American market. Bumble Bee's market share of U.S. tuna sales was 23 percent in 1991, up from 19 percent in 1990 (Handley 1991a).

Unicord sells tuna to the United States under its own brand labels and also sells under Bumble Bee through products generated at its canneries in California, Puerto Rico, and Ecuador. Despite their higher wage rates, the U.S. canneries were an advantage because they allowed Unicord to avoid high import duties of up to 12.5 percent on canned tuna. U.S. canneries were also relatively close to fishing grounds in the Atlantic and the ETP. The new technology of removing the loins, which are cooked, frozen, and shipped for canning within the United States, saved large amounts of money on shipping costs. Unicord was also in the process of setting up a loin operation in Ghana to serve the European market (Handley 1991b).

Unicord exports 40 percent of its Thai output to Europe. To avoid a 24 percent duty on canned tuna in Europe, Unicord was converting a herring cannery in Rostock, Germany, to can frozen loins. Dumri Konuntaklet hoped to eventually buy established brand names to gain an upmarket presence in various EEC counties and planned to enter the French market first (Handley 1991b). Although companies such as Thai Union have their own fishing fleets, Dumri said that practice is too capital-intensive and that having their own fleet would compete with the dominant fleets of South Korea, Taiwan, and Japan. "It creates a conflict of interest," Dumri Konuntaklet said. "We want to buy fish as cheap as we can" (Handley 1991b, 50). Unicord was rumored to be talking to McDonald's about a tuna product and already markets shrimp under Bumble Bee.

Like other Asian-based companies, Unicord, Southeast Asia's largest investor in the United States, is looking to the United States to secure its markets. Before the company bought Bumble Bee, it was the world's largest supplier of tuna but was at the mercy of industry middlemen, who bought the fish for resale to major brands. "Now Unicord can be assured of a distribution network in the United States, while Bumble Bee is sure of its supply," said Unicord chairman Kamchorn Sathirakul, a former governor of the Bank of Thailand. "Now we've become a truly integrated, global business" (Wallace 1992, H3).

Unicord's strong point has been low wages at its Thailand factory, where it employs 7,000 people to process raw tuna (Wallace 1992, H3). According to the USITC:

> Thai tuna processors enjoy a substantial labor cost advantage compared to U.S. processors. Thai wage rates are quite low compared with U.S. wage rates; a starting cannery worker in Thailand earns 90 baht (about $3.50) for an 8-hour workday, a minimum wage set by the Thai government. Several Thai industry sources reported that the average daily wage in tuna canneries is about 100 baht (about $3.90);

> non-wage labor costs for processors add another 30 baht (about $1.18) per worker per day, for a daily labor cost of about $5.10 per worker. Furthermore, minimum wages in remote areas are even lower and range from 74–84 baht per day ($2.90–$3.30). (1990, 77)

In 1990 there were 22 tuna canneries operating in Thailand. The top three firms account for 70 to 75 percent of the nation's canned tuna production: Unicord and Thai Union Manufacturing Company each account for about 15 to 30 percent, while Ta Kong Food Industries has a 10 to 15 percent market share (USITC 1990, 75). "Thai companies, especially in the food-processing business, are aggressively seeking out US companies which control their markets in order to lock up a foothold in fortress Europe and fortress USA," said Graham Catterwell, an analyst at Crosby Securities in Bangkok (Wallace 1992, H3; see also Burton 1989).

Heinz has strengthened its position in Puerto Rico and American Samoa and is pushing ahead with frozen loin operations but as yet has no processing in Europe (Handley 1991b). A supplement to Heinz's 1991 corporate report indicates that Heinz has Star Kist operations in California, Canada, Puerto Rico, France, Ghana, Ivory Coast, American Samoa (Pago Pago), and Australia (Heinz 1991). After being shut down since October 1985 because of a dispute over product quality and inspection, Star Kist Foods Canada reopened its St. Andrews facility in New Brunswick in November 1988. By mid-1990 they had again closed the plant, citing "price cuts by canners using Asian imports as the cause of the closure" (USITC 1990, 18).

Through the various acquisitions completed during this period, it is impossible to distinguish the national origin of tuna products. "Brand name recognition will continue to be important in the U.S. market, but the distinction between foreign and domestic canned tuna in the eyes of U.S. consumers and U.S. tuna wholesalers and retailers no longer exists" (National Research Council 1992, 32).

THE DEVELOPMENT OF TUNA LOIN PROCESSING TECHNOLOGY

The tuna loin is the lighter meat, similar to filets, and is produced by thawing, cooking, and cleaning frozen, whole tuna. The processes for producing tuna loins and canned tuna are the same up to the point where the tuna loin is produced. In loin production the loin is then packaged in plastic, frozen, and shipped to canneries, whereas in canned production the loin is packed directly into the can. The production of the loin accounts for about 80 percent of tuna canning labor costs. "By shifting production of loins to locations with relatively low labor costs, canned tuna producers realize substantial cost savings" (USITC 1992, 60).

In the 1960s, U.S. firms first experimented with importing frozen loins from Japan. In 1970, Bumble Bee, then owned by Castle and Cooke, purchased a tuna processing plant in Manta, Ecuador, as a transshipping facility to Puerto Rico. Shortly thereafter, the Ecuadoran government enacted legislation that required a certain amount of the fish to be processed in Ecuador. Bumble Bee then decided to produce a "relatively small quantity of loins for shipment to Puerto Rico in order to maintain the Ecuadoran facility" (USITC 1992, 60).

In the late 1970s, Star Kist contracted with tuna processors in Ensenada, Mexico, for frozen loins to be delivered by truck. Ensenada was Mexico's principal tuna production center and close to southern California. Van Camp also opened a cannery in San Diego in the late 1970s to import low-cost Mexican loins. This cannery used a state-of-the-art environmental control system designed to minimize odor and waste. The tuna embargo on Mexico in 1980 contributed to the closing of both plants (USITC 1992, 61).

The advantages of loins over the frozen whole tuna are varied. First, the labor savings are significant, since up to 80 percent of total labor costs in a traditional tuna cannery are employed in processing frozen, cooked tuna to the loin stage. Second, shipping costs are reduced, since only about one-half of the total fish weight is transported. Third, by only processing loins and not the whole fish the amount of waste is substantially decreased. This enables processors to streamline their production process and have "greater flexibility in choosing canning locations, since there is relatively little waste to dispose of when processing loins" (USITC 1992, 61). According to tuna industry sources, the relatively high cost of compliance with environmental regulations was also a significant factor in the closing of the mainland plants during the 1980s (USITC 1992, 61). Fourth, the use of loins enables canners to increase capacity with a minimal increase in capital investment and labor. Fifth, by processing only loins, firms can avoid the need to produce tuna-based by-products (e.g., cat food or fishmeal), as the market for these products has become increasingly competitive in recent years (USITC 1992, 61). Finally, another major advantage of the use of loin processing is that the duty on imported loins is much less than on canned imports. In 1991 the ad valorem equivalent duty rate on frozen tuna loins was 0.4 percent, while it was 11 percent on canned tuna (USITC 1992, 61).

There is one main disadvantage to loin processing—quality control. Loins are frozen and thawed one more time than the immediately canned raw product. This process often results in a less firm consistency of the tuna flesh as a result of ice crystal expansion. Although Star Kist and Bumble Bee argue that the differences between the quality of the frozen loins and frozen whole tuna are minor, Pan Pacific acknowledges that its loin operations yield a lower quality tuna product (USITC 1992, 62).

Loin processing is somewhat discouraged in American Samoa by the U.S. tariff on products of insular possessions—imported products that exceed 70 percent of the total value of the finished product. Currently, U.S. processors do not process loins in American Samoa, "mainly because of the relative abundance of raw whole tuna in the region as well as labor costs lower than California and Puerto Rico" (USITC 1992, 62).

As economic competition put pressure on U.S. processors to further reduce costs in the late 1980s, their interest in loin production resurfaced. For example, the new Bumble Bee plant in Sante Fe Springs, California, exclusively uses imported frozen loins, supplied mostly by its parent, Unicord. Other processors are reported to being considering such operations on the U.S. mainland. "The dolphin-safe policy announced by U.S. processors in early 1990 reinforced the advantages of using loins inasmuch as the primary source of raw material in the past for Puerto Rican plants, yellowfin supplied by U.S. vessels from the Eastern Tropical Pacific, was subsequently reduced" (USITC 1992, 62).

Unicord's purchase of Bumble Bee Seafoods accelerated Bumble Bee's use of tuna loins, and it is now the principal user of imported frozen loins in the United States. Tuna loins accounted for a substantial share of Bumble Bee's raw material input, and the firm processes tuna loins at both its Puerto Rico and Sante Fe Springs plants. The Puerto Rico operation gets its loins from the Manta, Ecuador, plant and the California plant from Thailand, mostly from Unicord. Bumble Bee has "relatively low production costs," "geographic proximity to sources of raw tuna," and possesses "superior canning technology, as well as the geographic proximity to major U.S. market areas" (USITC 1992, 62).

Star Kist has recently began using loins in Puerto Rico. The company's rather late entry is explained by several factors. First, unlike other U.S. companies that rely on American vessels for a high percentage of their tuna, Star Kist has contracted a larger share of its requirements from foreign fleets. Because of Star Kist's dominant size, it has the most sources of raw materials, and a better established global network reduces its costs. Second, Star Kist operates the world's first and second largest tuna processing plants in Puerto Rico and American Samoa, respectively. These two canneries provide all of the U.S. product and have to be operated at near capacity to be profitable. The costs of converting the canneries and retraining the reduced work force to process loins "have led to a reluctance by StarKist to shift a significant share of production to loin processing, which is still viewed as relatively high risk" (USITC 1992, 63).

Star Kist has increased loin production in Puerto Rico as a result of the decreased supplies from the ETP and the relatively high labor costs in Puerto Rico. Loin processing is not being pursued in American Samoa, where raw tuna supplies are abundant and labor costs are low. The company still retains the facilities in southern California that were closed in 1984. Although Star Kist's loin sources mainly have been the Latin American countries of Venezuela, Colombia, and Costa Rica, it currently procures loins from plants in Ecuador and Ghana. Star Kist terminated operations with Venezuela, Colombia, and Costa Rica because they lacked a dolphin-safe policy and were subject to the U.S. embargo (USITC 1992, 63; Woods 1992). The Ghana cannery began to send loins to Star Kist's Puerto Rico cannery in late 1990. Star Kist also recently purchased a fish cannery in Portugal, which could be used for loin production for processing in Puerto Rico but more likely will produce for the European market (USITC 1992, 63).

Caribe Tuna, a wholly owned subsidiary of Mitsubishi Corporation, operates a cannery in Puerto Rico that began to import loins from Ecuador in late 1990. According to Michael Dunn, the company's vice president, "Caribe made this change in response to rising labor costs in Puerto Rico as well as reduced availability of raw tuna from the ETP following the dolphin-safe policy in early 1990" (USITC 1992, 63). Caribe intends to increase its use of tuna loins at its processing site in Puerto Rico.

Van Camp Seafood does not currently use loin processing. Its cannery in American Samoa does not use loins for the same reasons as Star Kist. Van Camp's decision to close its Puerto Rico plant in 1990 indicated that it was not going to follow the same strategy as its competitors, increasing loin use in Puerto Rico to decrease labor costs. Van Camp is considering sending loins to Japan and Europe from its plants in Indonesia. According to a Van Camp Seafoods official, "Loins could be sent to Van Camp's plant in American Samoa in the future" (USITC 1992, 63).

Pan Pacific Fisheries, Inc., a wholly owned subsidiary of Marifarms, Inc., operates the last full-scale cannery on the continental United States. "Pan Pacific, which traditionally used raw tuna from its own vessels or from the relatively small local fleet, began to import frozen tuna loins in 1990, virtually all from Thailand" (USITC 1992, 64).

The labor effects of the extensive adoption of loin processing is significant, because a loin processing plant employs between one-fifth and one-third of the labor force of a full-scale cannery. If the whole U.S. industry converted, "cannery employment could fall by as much as 80 percent" (USITC 1992, 65). If the other processors follow Bumble Bee's lead, employment could decrease in Puerto Rico and American Samoa and relocate, although at only about 25 percent of the loss, in California (USITC 1992, 65).

Effects of Restructuring on the U.S. Tuna Industry

Tuna facilities are generally moving from Puerto Rico to American Samoa and on to Asian processing locations, closer to where the tuna are caught. Tuna caught by Asian fleets in the WTP and Indian Ocean do not associate with dolphins. Transnational tuna firms are bypassing the increased costs of fishing in the ETP by moving to the western Pacific, where regulations are minimal and labor costs are low. The final blow to the U.S.-based industry came when the three largest U.S. tuna processors—Heinz's Star Kist, Unicord's Bumble Bee, and Mantrust's Van Camp—decided in April 1990 to stop accepting ETP-caught tuna, i.e., dolphin-unsafe tuna. At the time, these three largest firms controlled about 80 percent of the U.S. market (USITC 1992, 9).

In 1991 the United States consumed about 600,000 tons of tuna, and the nation's seven tuna canneries employed about 10,000 people, mostly

Table 5.8. U.S. Canned Tuna Processors, Location by Firms and Processing Plants, 1991

Firm	U.S. Processing Plants
Bumble Bee Seafood, Inc. (Unicord, Thailand)	Mayaquez, Puerto Rico; Sante Fe Springs, California
Pan Pacific Fisheries (Marifarms, Inc., Terminal Island, California)	Terminal Island, California
Mitsubishi Foods Inc. (Caribe Tuna) Delmar, California (a subsidiary of Mitsubishi Corp., Tokyo, Japan)	Ponce, Puerto Rico
Neptune Packing Corp. White Plains, New York (a subsidiary of Mitsui USA, New York)	Mayaquez, Puerto Rico (inactive)
Star Kist Foods, Inc. California (a subsidiary of H. J. Heinz Company Pittsburgh, Pennsylvania	Mayaquez, Puerto Rico; Pago Pago, American Samoa; Terminal Island, California (inactive)
Van Camp Seafood Division (Mantrust, Indonesia)	Pago Pago, American Samoa; Ponce, Puerto Rico (inactive)

Source: Compiled from information submitted in response to questionnaires of the U.S. International Trade Commission as cited in USITC 1992, 9.

in Puerto Rico (Bradsher 1992). As already pointed out in the Introduction, the transnational move of the tuna fishing industry due to the largest processors' boycott (see Chapter 6 for an indepth discussion of the boycott) and associated high operating costs in the United States had important repercussions in terms of employment and overall economic well-being of fishing communities. The shift of production outside the United States, first to Latin America and then to Asia, involved a substantial growth in the demand for foreign products, mostly Asian. This situation translated into loss of employment in the United States and expansion of employment opportunities overseas.

In 1989 the bulk of the U.S. tuna processing industry was still located in Puerto Rico (see Tables 5.8 and 5.9). At the time, tuna companies operating in Puerto Rico were worried about a U.S. House bill that would allow twenty-two Caribbean Basin countries to export canned tuna to the United States at half the present duty. This action would further liberalize

President Reagan's Caribbean Basin initiative by reducing U.S. tariffs by 50 percent on select imports, including tuna, to provide encouragement for investment in the Caribbean. Randi Thomas of the Washington-based U.S. Tuna Foundation warned that passage of H.R. 1233 could encourage American and Asian companies to shift their operations from Puerto Rico, where they face federal minimum-wage laws and tough environmental regulations, to countries like the Dominican Republic, where wages are 30 cents an hour and environmental laws are lax (Luxner 1990).

Tim McCarthy, president of Bumble Bee Foods, which has 2,200 workers at its cannery in Mayaquez, Puerto Rico, said he was following the bill closely. According to McCarthy, Southeast Asia poses the greatest threat to the domestic tuna industry. According to the U.S. Tuna Foundation, in 1988 the United States imported 244.5 million pounds of tuna: 179 million pounds from Thailand, the rest from Taiwan, the Philippines, Ecuador, and others. In 1988–1989, Puerto Rico processed 57 percent of total U.S. cannery receipts (Luxner 1989).

The decision of the largest producers to boycott ETP tuna negatively impacted Puerto Rico. By 1991, 40 percent of the American-owned canneries in Puerto Rico had shut down (Kronman 1991). The Van Camp plant in Ponce closed in June 1990, and the Neptune Packing Plant, a subsidiary of Mitsui USA, closed its plant in Mayaguez in August 1990 (USITC 1992, 10). This left a total of seven U.S. tuna canneries (see Table 5.8). Many of the canneries that remained shifted some of their production to loin processing, which also decreases labor needs. From 1982 to July 1990, employment in Puerto Rico's tuna canneries declined from 15,000 persons to 6,600 persons; by the end of 1991 the number was 5,400. From 1985 to 1990 tuna processing employment rose from 3,318 to 4,700 in American Samoa. Employment at Pan Pacific's Terminal Island

Table 5.9. U.S. Tuna Canneries, by Plant Location, 1979–1991

	1979	1980	1983	1985	1989	1991
Continental United States	14	12	3	1	1	2
Hawaii	1	1	1			
Puerto Rico	5	5	5	5	5	3
American Samoa	2	2	2	2	2	2

Sources: 1979–1985 data from National Marine Fisheries Service as cited in USITC 1986, 172; 1989–1991 data compiled from data submitted to the U.S. International Trade Commission as cited in USITC 1992, 90.

Table 5.10. Average Number of Workers Employed in Establishments Producing Canned Tuna, Hours Worked, and Wages and Fringe Benefits, 1979, 1985, and 1991

	1979	1985	1991
Average number employed in the reporting establishments	15,831	14,197	10,898
All persons			
Production and related workers producing:			
All products	15,299	13,393	10,498
Canned tuna	14,668	12,887	9,613
Hours worked by production and related workers producing:			
All products (1,000 hours)	27,588	21,738	19,379
Canned tuna (1,000 hours)	25,661	21,121	17,934
Wages paid to production and related workers producing:			
All products (1,000 dollars)	119,774	106,362	96,044
Canned tuna (1,000 dollars)	110,741	101,745	86,916
Value of fringe benefits provided to production and related workers producing:			
All products (1,000 dollars)	24,220	13,630	24,892
Canned tuna (1,000 dollars)		13,037	20,663

Source: Compiled from data submitted in response to questionnaires of the U.S. International Trade Commission as cited in USITC 1986, 173 and 1992, 90.

plant fell from 1,228 in 1984 to 525 in 1989. In 1990 average hourly wages, including benefits, were $12.00 in the continental United States, $7.47 in Puerto Rico, and $3.40 in American Samoa. The average hourly wage per worker for all U.S. locations in 1990 was $4.77 (USITC 1990, 20–21; 1992, 10). By 1992, in Puerto Rico all but one of the five canneries had "either closed, cut back to a single shift, or made plans to do so by the time the canners' policy had been implemented" (USITC 1992, 42) (see Table 5.10).

By 1990, except for Star Kist's plant in Mayaguez, Puerto Rico, which employed 4,300 workers, the local tuna industry was virtually controlled by Asian firms: Unicord of Thailand, Mantrust of Indonesia, Mitsubishi and Matsui of Japan. For all of these firms, tax benefits under Section 936 of the U.S. Internal Revenue Code are crucial to their remaining in Puerto Rico (Luxner 1990).

In 1990, Bumble Bee in Puerto Rico employed 2,200 workers and processed between 200 and 300 tons of tuna a day, which accounted for more than 50 percent of Bumble Bee products sold in the U.S. mainland. "Bumble Bee started out in Astoria, Ore. We had plants in Hawaii, Maryland and San Diego, but currently operate tuna canneries only in Puerto Rico and Ecuador," according to Dan Sullivan, company president (Luxner 1990, 4A). Less than 5 percent of the production of Bumble Bee's Mayaquez plant went to the local market in Puerto Rico (Luxner 1990).

Boat owners in San Diego said that the decision of Heinz, Unicord, and Mantrust, the three largest tuna producers, to boycott ETP tuna was a tragic event that they had been fighting for twenty years, along with boat seizures, the closing of American tuna canneries, and foreign fleets slashing prices to capture the U.S. market. According to Peter Schmidt, president of Marco Seattle, whose Campbell Industries subsidiary in San Diego is one of the world's leading builders of purse-seine boats, "This [the boycott of ETP tuna] could be the last nail on the [coffin of the] American tuna boats" (Kraul 1990, D1). The last six canneries once located in San Diego closed in 1984, and local tunaboat owners "must now unload their fish at cannery plants in American Samoa and Puerto Rico" (Kraul 1990, D6). In the year following the large producers' boycott of ETP tuna, the number of U.S. fishing boats in the ETP dropped from thirty to nine, as tuna vessels relocated to the WTP, a shift that is inconvenient for U.S. processors who use Puerto Rico to process and can tuna (Wallace 1991). The dolphin-safe policy has made U.S. fleet access to the WTP fisheries critical (see Tables 5.11, 5.12, and 5.13).

Since the environmentalist victories of the 1990s, several boats in the U.S. tuna fleet, once the world's largest, have gone broke, while others were sold to foreign interests. Tunaboat captains had to relocate to the WTP and have shouldered $1 million to $2 million retrofits for larger nets, bigger hydraulics, and new engines (Kronman 1991). In 1990 in San Diego, there were forty giant boats, down from 130 in 1976, using purse seiners to fish in the ETP; by 1992 there were only nine left. This decline occurred as vessels were sold in response to the "dolphin-safe policy that forced the U.S. fleet to abandon the ETP. Since January 1990, 13 tuna vessels were sold to foreign fleets, 3 were operated by firms that went bankrupt, and 2 sank" (USITC 1992, 5). As of June 1992, only six U.S. vessels still fished the ETP; of these, two fish the smaller "school fish" that don't associate with dolphins, while the remaining four reportedly were continuing to "set on" dolphins and export their catch to Italy (USITC 1992, 5). "Over the past 35 years, the tuna industry's technology and

Table 5.11. U.S. Tuna Purse-Seine Fleet: Fleet Size, Additions During Year, Transfers to Foreign Flags, and Idle Vessels, 1978–1992 (capacity in short tons)

	Fleet size on January 1		Additions During Year		Transfer to Foreign Flag		Idle vessels on Dec. 31	
	No.	Capacity	No.	Capacity	No.	Capacity	No.	Capacity
1978	124	110,665	4	4,800	4	4,061		
1979	122	109,109	5	6,600	1	550		
1980	124	113,589	5	6,000	9	10,500	2	1,292
1981	117	107,349	13	14,750	1	550	11	8,587
1982	128	121,194	11	13,250	8	7,690	19	15,808
1983	125	124,173	6	7,750	5	5,081	34	28,933
1984	124	126,164			11	9,374	23	21,010
1985	107	110,985			9	8,585	21	22,422
1986	90	97,131	1	1,500	7	7,750		n/a
1987	80	87,899	4	3,800	12	12,110		n/a
1988	71	78,179	3	4,400	11	12,650		n/a
1989	63	69,929	3	3,700	2	2,400		n/a
1990	63	72,270	3	4,350	9	8,380		n/a
1991	56	67,140	2	2,850				n/a
1992	57	68,890			4	3,350		

Sources: Data through 1986 submitted by the American Tunaboat Association as cited in USITC 1986, 158; data from 1986 through 1992 submitted by the American Tunaboat Association, prehearing brief, July 27, 1990; correspondence dated Jan. 17, 1992, and Jan. 31, 1992, as cited in USITC 1992, 81.

capital investment has evolved around the giant purse-seiner boats, which can measure up to 220 feet in length, weigh 1,500 tons or more and cost $10 million each. There are about 380 of these ships operating in the world, of which 65 are registered in the United States," according to August Felando, president of the American Tunaboat Association (Kraul 1990, D6).

Within days of the U.S. boycott, prices paid to tuna fishermen dropped 22 percent to the lowest amount in ten years; yellowfin from the ETP fell from $1,075 per ton to $835 per ton (Kronman 1991). The shift of sixteen U.S. boats to the Western Pacific and abundant supplies of skipjack (75 percent) and yellowfin (25 percent) tuna increased yields and depressed prices. Dolphin-safe policies benefited newcomers on the tuna scene, notably Korea and Taiwan, who built boats and canneries as fast as they could. Those nations were already blessed with being close to waters that provide dolphin-safe tuna, not to mention lower overhead and regulatory costs than U.S. fisherman bear. The newcomers now have an artificial

Table 5.12. Number and Capacity of U.S. Baitboats and Purse Seiners Operating in the ETP, 1955–1985, (capacity in short tons)

	Baitboats		Purse Seiners		Total Fleet	
	Quantity	Capacity	Quantity	Capacity	Quantity	Capacity
1955	183	41,729	66	8,250	249	49,979
1960	80	15,691	111	24,971	191	40,662
1965	50	4,279	116	39,611	166	43,890
1970	45	3,852	118	55,823	163	59,675
1975	55	5,483	140	115,342	195	120,825
1980	25	2,186	117	107,734	142	109,535
1985	9	696	90	97,131	99	97,827
1986	3	348	64	43,235	67	43,583
1987	11	668	54	41,965	65	42,033
1988[a]	12	938	60	44,568	73	45,576
1989	9	839	51	33,009	60	33,848
1990	8	560	28	27,120	36	27,680
1991	8	560	17	16,590	25	17,150
1992			9	8,990	9	8,990

[a]Includes three jigboats.

Sources: Official statistics of the Inter-American Tropical Tuna Commission; submitted to the U.S. International Trade Commission by the American Tunaboat Association as cited in USITC 1986, 157 and 1992, 84.

advantage over the U.S. fleet, which must travel halfway around the world to catch marketable tuna (Kronman 1991).

As result of the boycott of tuna caught with purse-seine technology, the three largest U.S. tuna canners increased their reliance on Asian suppliers of canned tuna, such as the Philippines and Thailand. With a 65 percent market share, Thailand is the largest U.S. source of tuna, far ahead of Indonesia, the Philippines, and Taiwan (Luxner 1990)(see Table 5.14). U.S. tuna marketers increasingly are importing canned tuna to take advantage of lower labor costs in developing countries and to avoid dolphin-unsafe tuna (Thurston 1990). As a result, U.S. and foreign tuna firms are relying more and more on non–U.S. labor for processing. According to Randi Thomas, a representative of the Washington-based U.S. Tuna Foundation, "Most of the companies are not going to buy tuna unless they can assure that it was not caught in association with porpoises. It's extremely difficult to fish the ETP economically without porpoises. There will likely be more fishing in Asia and the Indian Ocean, which might be the new frontier" (Thurston 1990, 10a). The shift to other fishing regions hurt the volume of

Table 5.13. Location of U.S. Purse-Seine Fleet

Year as of Jan.1	Eastern Pacific		Western Pacific		Total	
	Boats (No.)	Capacity (Tons)	Boats (No.)	Capacity (Tons)	Boats (No.)	Capacity (Tons)
1990	28	27,120	35	45,250	63	72,370
1991	17	16,590	39	50,550	56	67,140
1992[a]	13	11,990	44	56,900	57	68,890

[a]Capacity includes two vessels with a total capacity of 1,700 tons not in operation in the Eastern Pacific and one vessel with a capacity of 1,400 tons not in operation in the Western Pacific.

Source: American Tunaboat Association as cited in USITC 1992, 83.

tuna caught by the U.S. fleet, which needed more access to other fishing areas. August Felando, president of the U.S. Tunaboat Owners Association, said that "they are trying to push us into other areas," as foreign suppliers capture a higher percentage of the yellowfin market (Thurston 1990, 10A).

When the ETP was profitable and legitimate, TNCs set up operations first on the U.S. mainland and then in Puerto Rico and Latin America to process for the U.S. market. When MMPA made the ETP illegitimate, processing facilities on the U.S. mainland and in Puerto Rico and Latin America became more costly and less convenient. The industry moved operations to the WTP, and processing centers shifted to American Samoa. As MMPA was increasingly enforced, production expanded for Asian processors and declined for U.S. and Latin American processors. The industry move to Asia to source dolphin-safe tuna and low-cost labor marginalized labor, both for tuna fishermen and tuna processing workers, in the United States (especially Puerto Rico) and in Latin America.

Summary: Global Economic Restructuring and Labor Marginalization

The events contained in the period from the late 1970s through early 1990s underscore the post-Fordist transformation of the tuna industry. Salient aspects of this transformation include the transnationalization of tuna production and processing. Tuna TNCs abandoned nationally based operation strategies in high-cost First World locations and globally sourced new, low-cost production sites as well as access to lucrative markets. The former objective was accomplished by the industry's eastward movement

Table 5.14. Canned Tuna: U.S. Imports for Consumption, by Sources, 1988–1991 (in thousands of kilograms)

Source	1988	1990	1991
Thailand	81,168	93,009	112,253
Indonesia	2,202	9,756	21,470
Taiwan	10,892	7,897	10,572
Philippines	8,394	12,268	9,956
Malaysia	1,281	1,333	1,900
Ecuador	3,773	1,540	1,183
Japan	1,529	639	417
Venezuela	80	1,036	795
Singapore	754	1,339	551
Spain	87	154	46
All others	748	691	407
World	110,907	129,090	159,550

Source: Compiled from official statistics of the U.S. Department of Commerce as cited in USITC 1992, 121.

across the Pacific Ocean. The penetration of lucrative markets was accomplished through acquisitions of U.S.-based firms, and the national identity of these acquired firms became a strategic tool in the hands of the tuna TNCs. Indeed, tuna brand names such as Chicken of the Sea and Bumble Bee continue to exist on the market, while their ownership and control moved away from First World–based companies. Post-Fordist transformation involved enhanced flexibility for TNCs. Episodes such as the reflagging of tuna boats and the ability to employ less expensive labor are cases in point. In the instance of labor, TNCs were able to move away from a high-paid Fordist labor force to low-paid post-Fordist labor. Consequences of this shift largely have been felt by communities such as fishing- and cannery-dependent locations in California, Puerto Rico, and other Latin American countries.

6
LAWSUITS, COMPROMISES, AND EMBARGOES

The senator's swelling chorus of praise is in fact a two-part harmony between the U.S. industry and politicians like himself. The U.S. industry invented purse seining on dolphins and for fifteen years monopolized the technique. The U.S. industry killed millions of dolphins in the early years of tuna seining, and in the seventeen years since MMPA was enacted, the U.S. industry has killed more than 800,000 dolphins. The U.S. industry has fought every regulation intended to reduce the dolphin kill. In 1980 an NMFS prohibition against "sundown" sets—implemented because the kill rate is up to four times as high at night as it is in daytime—was dropped, under pressure by U.S. industry lobbyists, after being in effect for just eight days. In 1981 the American Tunaboat Association sued to scrap the NMFS observer program. The observers' data, they argued, should not be used for enforcement. They won an injunction that kept all NMFS observers off U.S. tuna boats from 1981 to 1984, when the injunction was overturned on appeal. In the late 1970s, when forced to do so, the U.S. industry demonstrated considerable inventiveness in coming up with gear and techniques to minimize dolphin kills. That research has stalled, and the U.S. industry has done nothing favorable to dolphins lately. (Testimony of Sam LaBudde before the 1989 congressional hearings on MMPA reauthorization as quoted in Brower 1989, 58)

Attempts to Control the Foreign Fleet and Intimidation of NMFS Observers

Prompted by environmentalists' charges that foreign tuna fleets were still killing thousands of dolphins in the ETP, in 1984 Congress added two amendments to the original 1972 legislation to specify the standards and restrictions of the law (Davis 1988). The amendments stated that tuna caught using purse-seine nets in the ETP may only be imported if the government of the foreign country of origin demonstrated that it (1) had implemented a dolphin protection program "comparable" to that of the U.S. fleet, and (2) had an average incidental dolphin kill rate "comparable" to that of the U.S. fleet (Trachtman 1992). Congress ordered the NMFS to close the U.S. tuna market to nations failing to require dolphin protection measures comparable to those in the United States (Levin 1989).

In 1987 and 1988 numerous reports surfaced that crews of the U.S. tuna fleet forced observers to falsify dolphin kill numbers. Observers' lives were threatened, and seal bombs were thrown at them to make them leave their observation posts. One observer felt that if he fell off the boat he would be left to drown. Observers felt that they were exposed to "every possible form of harassment and coercion" (Anderson 1988). In addition, observers were only supposed to record the number killed, not dying or maimed, which consistently underestimated total dolphin mortalities. Much of the tuna fishing was done using "sundown sets" at night when observation was difficult (Anderson 1988). Three federal observers said that they were forced with threats and coercion to underreport dolphin kills and that federally reported numbers represented only half of total kills (*Audubon* 1988).

The U.S. General Accounting Office (GAO) looked into the NMFS program, including how observers were intimidated and numbers were underreported. The GAO probe was requested by members of the Senate Commerce Committee, chaired by Ernest Hollings (D-S.C.) and Pete Wilson (R-Calif.), in response to three affidavits from observers regarding intimidation and underreporting. NMFS said harassment was rare and did not lead to underreporting; if it did, the offenders would be prosecuted. In 1987–1988, there were eight cases of harassment or interference with fines, while in the previous ten years, there were only two cases of fines (Levin 1989).

1988—A Year of Turmoil

Although dolphin kills by the U.S. tuna fleet dropped through the 1980s, by 1988 the foreign tunaboats killed about four times as many dolphins as the U.S. fleet (*Audubon* 1988). According to Joshua Floum, a lawyer for Earth Island Institute (EII), by 1989 the foreign-flagged vessels were responsible for most of the dolphin kills in the ETP (*New York Times* 1989a, A17). "Many of the departed seiners have reflagged to avoid high U.S. operating costs and to escape the MMPA and other U.S. regulations" (Brower 1989, 57). Although Congress had ordered NMFS in 1984 to ban tuna imports from countries that did not have comparable dolphin kill ratios, these rules were not published until 1988 and gave other nations until 1991 to achieve the comparable kill rate. Frustrated by a lack of progress in the state arena, the environmentalists expanded their efforts to renew consumer boycotts of purse-seine tuna companies.

Stating that U.S. tuna companies had the "leverage to curtail the dolphin kill" by refusing to buy purse-seine-caught yellowfin tuna, in January 1988 a coalition of environmental groups launched a boycott of Ralston Purina's Chicken of the Sea, Heinz's Star Kist, and Pillsbury's Bumble Bee to pressure importers to take action. These firms processed about 70 percent of the tuna consumed in the United States (*Newsweek* 1990; Sharecoff 1990). The coalition was led by EII[1] and the Humane Society; they expanded the boycott in September to include pet food products (Davis 1988, 486).

At the same time, the environmental coalition accused NMFS of failing in its legal duty to protect the mammals and brought lawsuits in federal court in San Francisco (Levin 1989). "The federal government is supposed to ban the import of tuna from nations that ignore the quota, but has yet to do so," argued the environmentalists (*Audubon* 1988; 16). EII and the Marine Mammal Fund filed the lawsuit, which sought to force the U.S. Department of Commerce to impose a ban on imports from foreign violators and to properly enforce the porpoise quota on U.S. boats. The environmentalist coalition also urged Heinz and Ralston Purina to voluntarily end tuna purchases from nations that violated the quota (*Audubon* 1988).

In 1988 NMFS increasingly came under strong congressional criticism for delaying sanctions against foreign fleets that failed to reduce their dolphin kill. NMFS responded that any quicker action would have forced the foreign fleets to sell to other markets (Davis 1988). In 1987 a report by the Commerce Department's inspector general described the fisheries service's enforcement policies as lenient. "We were told by NMFS staff with longstanding and intimate knowledge of tuna fishing operations that fines have been so low compared to incomes that skippers have knowingly violated the regulations and accepted the fines," the inspector general said. Fisheries service officials responded that problems identified in the report were "real but transient" (Levin 1989, 35).

Environmentalists and industry officials differed over the definition of "comparable" kill rates and dolphin saving techniques as stipulated in the 1984 MMPA. Todd Steiner, a research biologist for Earth Island Institute, said that comparable "means the same." But Dave Burney of the U.S. Tuna Foundation said that "foreign countries must be brought along slowly so we don't lose them. If they choose not to export their tuna to the U.S., we lose a golden opportunity. We do not want to penalize the people who are making the effort out there. If we stay under 20,500 then everyone should be happy" (Godges 1988, 26). The Cetacean Society, Earth Island Institute, Greenpeace, Sierra Club, Whale Center, and other groups pooled their resources as the Marine Mammal Protection Act Reauthorization Coalition to push for needed improvements in the law and at the same time to ask for consumer boycotts (Godges 1988). The National Audubon Society also joined with EII and the other environmentalist organizations at the MMPA reauthorization hearings to try to amend the 1984 law to reduce the dolphin-kill quota to zero (*Audubon* 1988).

Crucial testimony at the 1988 reauthorization hearings in support of the environmentalist agenda was provided by Sam LaBudde. For five months in 1988, LaBudde worked as a cook on a Panamanian tunaboat and documented the mass killing of dolphins. "I saw hundreds of animals being drowned, mutilated and butchered," LaBudde said. "Many of the animals were caught in the mesh and hoisted out of the water. Some will fall back into the sea as flippers and beaks are broken or ripped out of their bodies only to become ensnared moments later and pulled out once again" (Davis 1988, 488). He told of one case where more than 200 dolphins were slowly drowned to catch less than a dozen tuna (Davis 1988).

EII sponsored LaBudde's investigative work. An eleven-minute edited version of the video "where dolphins squealed in pain as they succumbed—in some cases being ground up alive in the gears of the nets—was first aired in March of 1988, to horrified audiences" (Kraul 1990, D6). The video made the issue of dolphin killing terribly real to millions of Americans (Kraul 1990). EII also distributed the film in Europe to try to stop imports from the ETP into Spain, Italy, France, and Portugal (Anderson 1988). After the release of the LaBudde film in Europe, a Greenpeace-sponsored motion to ban tuna imports from countries that net dolphins was supported by all parties of the European Parliament (Davis 1988). When LaBudde showed his film before the congressional hearings on MMPA in 1988, a chorus of senators applauded the U.S. industry and lambasted the foreign fleets.

At the 1988 reauthorization hearings the environmentalists asked for a four-year phaseout of purse-seine fishing but did not get it. "We had practically the entire environmental community back at the authorization hearings, everyone from Audubon to the Humane Society," LaBudde said. "Twenty-eight national environmental organizations wanted purse-seining stopped. Eliminated. We asked for a four year phase out. That would give the Marine Mammal Protection Act twenty years to do what it was designed to do—reduce kills to insignificant levels approaching zero. We thought four more years was a reasonable time. We got beat by owners of thirty-five tuna boats" (Brower 1989, 58).

Although environmentalists did not get their mandated phaseout of purse-seine fishing, the 1988 amendments to MMPA did require U.S. boats to have a special panel to release the dolphins. Fishermen also were required to enter the water in rubber boats or wetsuits to help release the dolphins, but still many dolphins died—although a netted catch often did work without killing any. Industry spokespersons reported that 99 percent of dolphins were released, but environmentalists still asserted that the average yearly U.S. kill of dolphins was 18,000, even though the U.S. fleet has dropped from ninety boats in 1981 to about thirty-five in 1990. According to the environmentalists, "The reason is they have been catching more yellowfin and relying more on herding dolphins to do it" (Levin 1989, 36).

Environmentalist lobbying at the 1988 MMPA reauthorization hearings produced amendments and implementing regulations with specific requirements that every country expecting to export yellowfin tuna from the ETP to the United States was required to meet. These requirements included

> participation in the observer program of the Inter-American Tropical Tuna Commission (IATTC); that vessels execute a proper backdown procedure to release dolphins; that each foreign vessel be

> equipped with a dolphin safety panel (Medina panel) to prevent the entanglement of dolphins during the backdown; that sundown sets and the use of explosives to drive dolphins be prohibited; that vessels have on board at least three speedboats equipped with bridles, towing lines, and snap hooks to prevent the collapse of the net; that vessels be equipped with a platform and underwater observation gear to be used for the observation and rescue of dolphin; and, finally, that vessels be equipped with long-range floodlights to be used in case the backdown channel has to be illuminated to direct the release of dolphins. (Colson 1992)

The 1988 law also clarified the meaning of comparable kill ratios with provisions requiring foreign fleets to prove that their dolphin kill rates were no more than twice the U.S. rate of 20,500 for the year 1989. The new law required NMFS to determine whether the foreign fleet had complied by the end of 1989 and whether their kill rate was no more than 1.25 times that of the U.S. fleet's rate in 1990 (Morain 1990).

LaBudde's testimony critically illustrates how MMPA regulations were challenged at every juncture by the tuna industry and fractions of the U.S. state. Environmentalists had to use the courts and sympathetic legislators to force compliance from NMFS and the Department of Commerce. At the MMPA reauthorization hearings in 1988, several U.S. senators expressed displeasure with NMFS and its parent, the National Oceanic and Atmospheric Administration, "for their failure to implement the regulations that would keep that can off the shelf." Sen. Bob Kerry (D-Nebr.) pointed out that MMPA was amended in 1984 to require foreign nations to meet comparable U.S. dolphin kill rates or face a ban on imports. "Why had the NMFS taken four years to formulate 'interim final regulations' to that end?" (Brower 1989, 58). According to Charles Fullerton of NMFS, "It's a very delicate operation to get those regulations. We developed some over a year ago which were not acceptable either to the tuna industry or to the foreign nations. So we went back to the drawing board and developed a whole new set, the ones that are now in interim phase. We'd like to give these a try" (Brower 1989, 58). Brower writes, "How could a bureaucrat in a regulatory agency so lose track of his mission? The proposed NMFS regulations were not acceptable to the tuna industry or the foreign nations—the regulatees—so of course the regulators scrapped them" (Brower 1989, 58).

FEDERAL JUDGE ORDERS MORE OBSERVERS

In January 1989, a coalition of environmental groups filed lawsuits against the U.S. government, claiming that the presence of observers on only one-third of U.S. tuna boats was inadequate. NMFS said there was not enough funding to put observers on each boat, but the judge rejected those arguments as well as arguments by the American Tunaboat Association that the failure to station an observer aboard every vessel would not cause "catastrophic consequences" (*New York Times* 1989a, A17).

In August 1989 Judge Henderson ordered observers on all tunaboats that sell tuna to the U.S. market. The American government and the U.S. tuna industry told Judge Henderson that U.S. trade would be hurt and few dolphins saved if observers were required to monitor every foreign tunaboat to reduce dolphin killings. Joshua Floum, a lawyer representing EII and the Marine Mammal Fund, responded that U.S. law requires that foreign-owned tunaboats selling their catches in the United States meet the same standard as boats owned by U.S. fishermen (*New York Times* 1989b, A14). The government and the tuna industry disagreed, arguing that the law allowed the Commerce Department to set the number of observers on foreign boats and that the number had been set at 33 percent. David Burney, U.S. Tuna Foundation spokesperson, said that "if we required 100% of foreign boats to have observers it would drive the foreign boats to sell their catches to nations who don't regulate dolphin killing" (*New York Times* 1989b, A14).

The Inter-American Tropical Tuna Commission (IATTC) was called upon more and more to help resolve the growing tuna-dolphin controversy. In 1989, all countries with purse seiners in ETP dolphin waters participated in the IATTC program that puts observers aboard one of every three tunaboats. Observers provided data to estimate "incidental mortality of dolphins," said Martin Hull, IATTC tuna-dolphin expert (Tennesen 1989, 12). IATTC also mediates territorial fishing water disputes. During the 1950s and 1960s, many U.S. boats were seized by foreign governments over territorial water disputes, and the U.S. State Department challenged the extension of the 200-mile national limits. Some countries have entered into agreement (the United States, Panama, Honduras, and Costa Rica) to internationalize tuna fishing rights (Tennesen 1989). "Tuna force people to think internationally," says August Felando, president of the American Tunaboat Association (Tennesen 1989, 13). In a world hungry for protein, fisheries experts say tuna needs more regulation, but fishermen oppose such regulation. "Nations have to act cooperatively to ensure a continuing supply of tuna," said James Joseph, IATTC director of investigations (Tennesen 1989, 13). Tuna is a fish without national boundaries, and in this case, man has to adjust his laws to nature's laws (Tennesen 1989).

Dolphin-Safe Labels and the Dolphin Protection Consumer Information Act

The bypassing of the action of the U.S. state by the tuna industry generated responses from the environmental groups that supported MMPA. Having failed to win legislation mandating a phaseout of purse-seine netting at the 1988 MMPA reauthorization hearings, late in 1989 environmental and animal rights groups again renewed their attack at the consumer level. Environmentalists proposed the Dolphin Protection Consumer Information Act, sponsored by Rep. Barbara Boxer (D-Calif.) and Sen. Joseph Biden, Jr. (D-Del.) along with 100 other sponsors, to counter trends toward purse-seine nets. Advocates of the bill wanted product labels stating that "The tuna in this product has been captured with technologies that are known to kill dolphins" and also a label stating that the product was "dolphin-safe" (Salmans 1990, 84). Environmentalists argued that supermarket chains such as Sainsburys of England and Loblaws of Canada already carried such labels (Salmans 1990).

NMFS director Charles Fullerton criticized the proposed act, saying that only 10 percent of the U.S. fleet's tuna catches involved dolphin catches, but that 70 percent of foreign yellowfin tuna catches did (Vickers 1989, 27). The U.S. tuna industry argued that the law was harmful to U.S. processors. According to David Burney of the U.S. Tuna Foundation, "The law would be a logistical nightmare . . . animal-protection groups have exaggerated the number of dolphin kills" (Vickers 1989, 27). August Felando, president of the American Tunaboat Association, said the bill represented abuse of congressional regulatory powers (Vickers 1989). The Bush administration opposed the bill, saying that the "voluntary policy adopted by the canners will allow market forces to solve the problem" (Taylor 1990, 1553).

The Dolphin Protection Consumer Information Act was passed in 1990 mandating standards for the labeling of tuna cans as "dolphin-safe." It also required the Secretary of State to "immediately seek, through negotiations and discussion with appropriate foreign governments, to reduce and, as soon as possible, eliminate the practice of harvesting tuna through the use of purse seine nets intentionally deployed to encircled dolphins" (National Research Council 1992, 33).

Big 3 Tuna Canners Boycott ETP Purse-Seine Tuna

After years of pressure from environmentalists and faced with a growing consumer boycott of tuna, in April 1990 the U.S. tuna industry finally changed its tuna fishing policy. Heinz, Van Kamp, and Bumble Bee—which accounted for 70 percent of canned tuna sold in the United States—announced that they would no longer accept tuna that was caught in nets that kill dolphins. The ban included purse-seine nets and drift nets. First Heinz, the world's largest tuna firm, announced the decision, and then the other two companies followed suit. A Star Kist representative said that there were other ways to catch tuna. Heinz Chairman J. F. O'Reilly said his company was responding to consumer concerns, including postcards and letters from schoolchildren, in adopting a policy of "dolphin-safe" tuna and hoped that this would increase sales. Heinz said it would rely on reports from NMFS observers to assure dolphin-safe tuna (*Newsweek* 1990).

Environmentalists had pressured the tuna industry for years, with tactics that included a boycott of canned tuna begun in the early 1970s as well as over 7,000 telegrams (Kraul 1990). David Phillips, director of EII, said, "If fully implemented, Heinz's actions will represent the most important step for protecting dolphins since the passage of the MMPA" (Sharecoff 1990, A14). Phillips also said that O'Reilly and other Heinz executives indicated they would join the environmental community in pressing for legislation that bars the sale of tuna caught in association with dolphins (Sharecoff 1990). "Tuna is fun food," said Heinz Vice President Ted Smyth. "If it's associated with the harassment and killing of a noble creature like the dolphin, that's not right" (*Newsweek* 1990, 76). All three firms agreed to put a "dolphin-safe" logo on their tuna. The catalyst for the decision by Heinz was a lunch meeting between Jerry Ross, chairman of A and M Records, and J. F. O'Reilly, chairman of Heinz. Over lunch they discussed the LaBudde video. Ross told O'Reilly that reduced kills were not enough and that "people just want to let the dolphin alone, period, and they're willing to pay the extra cost to do that" (Parrish 1990b, D1).

Felando, of the American Tunaboat Association, who represented the twenty-nine boats of the San Diego tuna fleet, said that the Heinz action was merely a marketing ploy to increase sales and that it would force fishermen to catch younger tuna, which eventually will lower the population (Sharecoff 1990). Felando also said that the decision by the large companies "would only serve to penalize the [U.S.] fishing fleet, which has improved its methods for protecting dolphins" (*Time* 1990, 63).

Dolphin kills were down significantly because of NMFS observers aboard U.S. tunaboats but were not low enough to satisfy the environmentalists. Under MMPA all U.S. boats had NMFS observers and one-third of the foreign boats had IATTC observers, but environmentalists said NMFS did not enforce the law on foreign boats (Sharecoff 1990). U.S. boats could go farther out to sea, where tuna do not associate with dolphins, but in general the U.S. boats were too small. Fishing waters in the Western Pacific were "scenes of intense competition among Asian, French, and Spanish boats" (Kraul 1990, D6). Tuna industry officials were also skeptical that the U.S. canners' policy would significantly reduce dolphin kills and said that this would only allow foreign fleets to catch the easy tuna in the ETP without NMFS observers and then sell it elsewhere or transship it into the United States (Kraul 1990).

Bumble Bee Accepts Dolphin-Unsafe Tuna

In late 1990, environmentalists claimed that Bumble Bee lied when it said it would no longer accept tuna caught with dolphins and promised to sell only dolphin-safe tuna internationally. EII reported they had proof that Bumble Bee accepted a shipment of dolphin-killing tuna in Thailand and argued that Bumble Bee still used coastal drift nets, which kill dolphins. Save the Dolphin Project Associate Director Brenda Killian said she was told by members of Unicord they had accepted a shipment of dolphin-unsafe tuna that had been refused by other buyers. Bumble Bee first said that the buyers were not actually Unicord companies (Parrish 1990b, D2) and threatened to sue EII for libel. Phillips of EII said Unicord was persistently buying dolphin-unsafe tuna and urged a boycott. Bumble Bee countered that it agreed to a phaseout program to honor existing contracts of fishermen but that only dolphin-safe tuna had been sold in the United States since the agreement (Meier 1990b). Bumble Bee later admitted that it did buy the dolphin-unsafe tuna without checking the observer documents but that it was a mistake (Meier 1990a).[2] In response to environmental attacks, in May 1991 Bumble Bee gave $500,000 to NMFS and IATTC to study how to catch tuna in the ETP without catching dolphins. Unicord also accepted an EII team in Thailand, including a full-time EII inspector to work from Unicord offices in Thailand, and provided full access to company documents regarding tuna fishing and dolphin kills (Handley 1991a).

In an attempt to monitor and regulate the global tuna fishing industry, in December 1990 Greenpeace and the Dolphin Coalition drafted a five-point corporate policy that they wanted the international tuna-packing industry to adopt. The corporate policy stated that for a canner to claim to be selling only dolphin-safe tuna, the policy "must be binding worldwide, including all subsidiaries, controlled bodies . . . enterprises which purchase, process or sell canned tuna or tuna products for export" (Parrish 1990a, D4). Gill net fishing must be stopped and companies must provide access to all company records regarding tuna buying. The plan focused on the international aspects of tuna fishing. "The Italian canneries are buying up this cheap yellowfin tuna set on dolphins," said Brenda Killian of EII (Parrish 1990a, D4). EII also questioned the practices of Mitsubishi, Mitsui and Company, and Bumble Bee, in effect calling for an international accord to monitor the tuna fishing industry, the role a transnational state would perform.

The First Embargo and Calls for an International Forum

In August 1990, EII and other plaintiffs brought suit in the U.S. District Court for the Northern District of California, seeking to enjoin the Secretary of the Treasury and other defendants to take action to implement the import restrictions of MMPA. Until required in 1990, NMFS had failed to make available comparability findings on mortality, which were necessary to implement the 1988 amendments, and thus allowed imports of tuna regardless of foreign dolphin kill rates (Trachtman 1992). This failure of NMFS to produce the comparability data was a result of the lack of consensus on the exact interpretation of MMPA's intent. The NMFS interpretation of MMPA centered on the maintainence of optimum sustainable populations (OSP), while the environmentalists' view focused on reducing dolphin kills to "insignificant levels approaching 0." This unresolved controversy hampered NMFS from providing timely data, which triggered the ensuing lawsuits.

Soon after the suit was brought, Judge Henderson ordered the Bush administration to impose an immediate embargo on imports of tuna caught by foreign fleets until they proved they were reducing the number of dolphins killed. This action mostly affected Mexico, Venezuela, Panama, Ecuador, and Vanuatu (Morain 1990). Judge Henderson said that the Bush administration was taking too long in determining whether foreign fleets were complying with U.S. law (*New York Times* 1990). He also charged NMFS with not enforcing a 1988 MMPA provision that ordered foreign fleets to prove they were reducing their dolphin kills to comparable U.S. levels. "Simply put, the continued slaughter and destruction of these innocent victims of the economics of fishing constitute an irreparable injury to us all, and certainly to the mammals whom Congress intended to protect," stated Henderson (Morain 1990, A3).

"This is a stunning rebuke for the government's position," said David Phillips, director of EII, which sued to force the federal government to take action against the foreign nations (Morain 1990, A3). "Basically, the State Department has been categorically against enforcing these embargoes," Phillips continued. "They have put dolphin protection at the bottom of the priority list in dealing with these countries" (Morain 1990, A19). Judge Henderson accused the foreign nations and the U.S. government of "foot-dragging." The embargo mostly affected Mexico, which exported about 20,000 tons of yellowfin tuna per year to the United States. The Pacific Ocean island of Vanuatu exported 72 percent and Venezuela 48 percent of their catches to the United States at the time (Morain 1990).

U.S. dolphin kills in the ETP were down dramatically, but Mexico still had fifty ships using purse-seine methods there. The U.S. dolphin kill was 12,643 in 1989, down 36 percent, but NMFS said 1988 foreign fleet kills were up 40 percent, to 84,000. Judge Henderson said that the failure of Commerce Secretary Robert Mosbacher and the National Oceanic and Atmospheric Administration to enforce the 1988 provision "assures the continued slaughter of dolphins" (Morain 1990, A19). "The statute was intended to use access to the U.S. market as an incentive for foreign nations to reduce marine mammal deaths. The Secretary, contrary to Congressional intent, has not provided that incentive," wrote Henderson (Morain 1990, A19). The Commerce Department disagreed with Judge Henderson, arguing that the 1988 amendments to MMPA only required the Commerce Department to collect information on foreign fleet dolphin kills, not to initiate embargoes (*New York Times* 1990).

Although in August 1990 the district court found in favor of the plaintiffs and ordered the Secretary of the Treasury to impose embargoes on imports from Mexico, Venezuela, Vanuatu, Panama, and Ecuador, the next day NMFS made positive findings for Mexico, Venezuela, and Vanuatu, allowing the embargo to be lifted on them. Panama and Ecuador were later exempted after they prohibited their fleets from setting on dolphins (Trachtman 1992). Asserting that NMFS had not correctly figured their dolphin kill rates for Mexico, EII sought a restraining order on Mexico. In October 1990 the district court granted the temporary restraining order and converted it into a preliminary injunction reinstating the embargo on Mexico. The U.S. government appealed the injunction, arguing that it was the government's discretion (Commerce Department) to interpret MMPA. The U.S. Court of Appeals found in favor of the Commerce Department and removed the embargo on November 14. In February 1991, the appellate court vacated the stay of the appeals court and reinstated the embargo. In March the embargo was extended to include Venezuela and Vanuatu. On April 11, 1991, the appellate court held that NMFS's interpretation conflicted with statutory language and congressional purpose (Trachtman 1992). The U.S. Department of Commerce lost its appeal.

The embargo was enforced reluctantly by the Bush administration, which insisted that Mexico was making progress in reducing dolphin kills. The measure also prompted a bitter denunciation by Mexico of an example of American coercion. The U.S. government and environmentalist officials said that for offending countries like Mexico, which exported $56 million in tuna last year, the costs of the tuna conflict would grow dramatically if no attempt was made to abide by U.S. restrictions (Uhlig 1991).

The embargo had only a limited effect on U.S. tuna imports, since the three largest U.S. tuna firms already bought only dolphin-safe tuna.

Under provisions of the law, the embargo would automatically be extended after ninety days to tuna imports from all intermediary countries that buy and process tuna from offending fishing fleets. After six months, experts said, the embargo may be extended to cover other fish products from violating countries, potentially affecting some $400 million in other Mexican fish exports such as shrimp (Uhlig 1991).

As part of revised 1988 MMPA, Mexico was the only nation that exceeded the quota of dolphin kills, according to U.S. figures. The embargo prompted accusations of U.S. protectionism. According to Mexico's Secretary of Foreign Relations, "In accordance with international law, no country has the right to impose their own criteria on others, much less apply sanctions" (Scott 1991b, 6). Mexico reduced its dolphin kills by 67.5 percent between 1986 and 1990, according to a Mexican Fishery Secretariat study, which noted that the U.S. tuna fishing industry had fifteen years to make the same reduction. Several prominent Mexican politicians and business leaders were irked by what they saw as a ploy to protect the U.S. market share by forcing a poor, developing nation to meet unreasonably high ecological standards. They wanted this dispute and the issue of environmental trade barriers on the table in the free-trade talks. "This is a particularly severe warning for the free-trade agreement negotiators of the loop-holes to watch out for," said Hermenegildo Anguianos, a congressman belonging to the Institutional Revolutionary Party, or PRI (Scott 1991b, 6).

A spokesperson for President Salinas de Gortari said the official Mexican position was that the disagreement on the tuna issue should be handled outside the free-trade talks. Both Salinas and Bush administration officials agreed that standards for protecting dolphins should be set in international forums, as has been done for whales. But unless the U.S. law was changed by Congress, State Department officials said the embargo was likely to stand. U.S. sanctions on Mexico were an embarrassment to the Bush administration, which was trying to get congressional approval of fast-track negotiations on the North American Free Trade Agreement (NAFTA) (Scott 1991a).

Initially the embargo had little economic effect on Mexico. In 1990 sales to the United States only accounted for about 3 percent of tuna exports, and most Mexican tuna was sold to Europe and Japan. But on April 20, 1991, a secondary embargo went into effect that significantly curtailed the volume and prices of Mexican tuna exports. The United

States, which is the largest consumer market for tuna, extended the ban to all countries that bought Mexican tuna to can and reexport it to the United States (Scott 1991a). "We have nothing against Mexico. It just has to stop killing dolphins," said Joshua Floum, attorney for EII. "The writing is on the wall. Markets are closing around the world to 'dolphin-unsafe tuna' " (Scott 1991a, 6).

MEXICO, GATT, AND NAFTA

On November 5, 1990, Mexico requested consultations with the United States regarding the tuna embargo. Consultations were held December 19, but a satisfactory solution was not reached. Faced with further losses or potential sanctions for its $450 million export fishing sector, Mexico began proceedings against the United States at the General Agreement on Trade and Tariffs (GATT) (Uhlig 1991). In January 1991 Mexico requested a panel before GATT to present its position on the U.S. tuna embargo. GATT agreed and the panel was formed March 12, 1991 (Trachtman 1992).

The GATT hearings occurred when Mexico and the United States were trying to negotiate the North American Free Trade Agreement (NAFTA). With NAFTA at a critical stage, the United States and Mexico were locked in dispute over the environmental damage caused by Mexico's tuna fleet. With growing international support for dolphin-safe tuna and for foreign fleets such as Japan, Taiwan, and South Korea to change their fishing practices, the issue threatened the Bush administration with a volatile trade battle with Mexico at the very moment it was trying to court and defend Mexico as a major trading partner. Mexican critics of the free trade accord seized upon the conflict as an example of American domination under any such pact. American conservation groups cited the behavior of the Mexican tuna fleet to emphasize the need for tough environmental scrutiny of all aspects of a free trade accord (Uhlig 1991). According to environmentalists, even with sanctions and embargoes, foreign fleets in Mexico, Venezuela, Vanuatu, and Ecuador continued to use the "dolphin set on" method. "In terms of sheer numbers, Mexico kills the most dolphins of any country by far," said David Phillips of EII (Uhlig 1991, D2).

The GATT panel found in favor of Mexico and stated that "regulations governing the taking of dolphins incidental to the taking of tuna could not possibly affect tuna as a product" (Trachtman 1992, 146). Environmental concerns about the method of production, as opposed to the product itself, are excluded from consideration under Article III of GATT, as are, for example, "concerns about the treatment or human rights of workers" (Trachtman 1992, 147). In essence, countries cannot embargo a product for how it is produced.

A three-man committee ruled that the U.S. law was contrary to GATT's equal treatment provisions. Moreover, the committee drew conclusions that went beyond the dolphin case. In its confidential (but quickly leaked) report, it concluded that "a contracting party may not restrict imports of a product merely because it originates in a country with environmental policies different from its own" (*Economist* 1991, 31). The GATT ruling threw into doubt all kinds of environmental laws that impose restrictions or penalties on foreign countries (*Economist* 1991). Environmentalists argued that a GATT ruling in favor of Mexico could set a precedent that might undermine their efforts on a range of fronts. "If the GATT ruling goes through, international trade sanctions designed to halt trade of endangered species, trade in rare hardwoods, and shipments of toxic wastes could be declared illegal. It would be a very serious blow," said a spokesperson for EII (Scott 1991a, 8).

Critics of the embargo argued that the rule stating that foreign dolphin kills can only be 125 percent of U.S. dolphin kills was hard to interpret. The rule was based on the actual kill of the U.S. fleet, and not 20,500. Only after the tuna were caught and the dolphin kills recorded by the U.S. tuna fleet did the actual number become published. Competition between Latin American and U.S. tunaboats is highest in the ETP. But in the Western Pacific, where U.S. fleets can go, Latin American boats did not have the same access to ports or fishing rights. Mexico was suspicious that the law was designed to protect U.S. fishermen, not dolphins, and raised the issue of whether the United States has the right according to GATT to unilaterally enforce a limit on Third World countries (*Economist* 1991).

Resource-rich developing countries saw imperialism in the efforts of industrialized nations to dictate environmental reforms and raised the charge of trade protectionism masquerading as concern over the environment. In the United States, environmentalists saw a threatening side to the Bush administration's efforts regarding NAFTA and feared that U.S. environmental laws would be compromised in the process (Magnusson et al. 1992, 130). "The administration is working behind the scenes to achieve some of the deregulation that it was not able to get in the open," charged Lori Wallach, a lawyer at Congress Watch, a group founded by Ralph Nader (Magnusson et al. 1992, 130). The U.S. Congress, especially Reps. Henry Waxman (D-Calif.) and Richard Gephardt (D-Mo.), were strongly opposed to amending MMPA to suit the federal government and the U.S. tuna industry, or to any Mexican deal that imperiled U.S. health, safety, labor, or environmental laws (Magnusson et al. 1992).

According to William K. Reilly, head of the Environmental Protection Agency, "If this becomes the basis of GATT policy, it would unravel all the strings" of U.S. environmental policy (Magnusson et al. 1992, 130). If the GATT decisions prevailed, U.S. trade officials said the effects might also weaken enforcement of international environmental accords, e.g., sea turtles, ozone, rain forests, endangered species, whaling, ivory, and elephants. According to GATT, such problems concerning the "global commons" should be solved through "international environmental agreements" (Brooke 1992a, 7). GATT even called for an international forum to resolve "global commons" issues such as MMPA.

THE FIRST COMPROMISE: THE UNITED STATES, MEXICO, AND GATT

In September 1991 U.S. officials reached an understanding with Mexico over the U.S. embargo on Mexican tuna and the GATT decision. "Guillermo Jimenez Morales, Mexico's fisheries minister, told reporters that his nation would defer action on its GATT complaint, in exchange for a promise from U.S. officials to lobby Congress for changes to the dolphin-protection law, the MMPA" (Senzek and Maggs 1991, 1A). Morales said the proposed changes would have the effect of lifting the tuna embargo. U.S. State Department officials portrayed the commitment as simply following through on a previous plan to change the law to allow Mexico and other countries various options for meeting its requirements and thus avoiding the embargo. "We did not agree to anything new that I am aware of," said a U.S. official (Senzek and Maggs 1991, 1A).

The compromise struck between the United States and Mexico included Mexico's promise to change fishing styles and the United States' promise to seek changes in MMPA that would lift the embargo on Mexico, including lobbying environmental groups and Congress by arguing that lifting the tuna ban would provide an interim solution that would save more dolphins in the short term (*Journal of Commerce and Commercial* 1991). "The promise to try to amend the law was brokered in Mexico two weeks ago by Secretary of State James Baker, Secretary of Commerce Robert Mosbacher and U.S. Trade Representative Carla Hills," according to Joshua Bolten, Hill's general counsel (Maggs 1991, 3A).

In exchange for the Bush administration's pledge to try to change MMPA, Mexico issued a ten-point plan to reduce dolphin kills by fishermen and deferred action on the GATT ruling. In September 1991, President Salinas announced that, as "a show of good faith," Mexico would "postpone" the final GATT decision and pursue bilateral solutions. Fighting for the Mexican tuna industry was one thing, but being cast as a nation that was downgrading environmental efforts may have scuttled the NAFTA talks (Scott 1991a). In October 1991, Salinas flew to California to announce his ten-point plan (*Economist* 1991). To counter the possible negative effects of the GATT ruling on American public opinion during the NAFTA talks, Mexico also hired a Washington, D.C., public relations firm to place full-page ads in the *New York Times* and other prominent American newspapers, assuring the United States that it was pursuing dolphin-friendly policies. EII called Salinas's proposal cosmetic (Scott 1991a).

The proposed changes were transmitted to the Senate early in 1991. Amending the law faced overwhelming opposition on Capitol Hill. Even if Mexico dropped its complaint, Subcommittee Chairman Henry Waxman said that he was worried the GATT ruling indicated that many other environmental laws would be challenged under the agreement. According to Waxman, "The decision places a cloud over a large number of environmental laws that use trade sanctions to protect whales, elephants, baby seals, and endangered species, as well as our recent ban on drift-net fishing. Also at risk is the Montreal Protocol . . . designed to protect the ozone layer" (Maggs 1991, 3A). Waxman and other members of Congress urged the Office of the U.S. Trade Representative to try to change GATT rules that were used to throw out the embargo. According to Timothy O'Leary, a spokesman for that office, "We argued very vigorously for the ban before the GATT panel this fall. It gets kind of touchy, but a better way to have gone about protecting the dolphin would have been some kind of international convention like the ones to protect the ivory or ozone, rather than one nation taking unilateral action" (Parrish 1991, D1).

According to some U.S. officials, any attempt to change the law, short of major new commitments from Mexico to reduce dolphin kill rates in tuna fishing, would rile wildlife protection groups that had already won a round in court against the Bush administration over its enforcement (Senzek and Maggs 1991). Critics also argued that attempts to weaken MMPA would attract the attention of hard-core opponents to NAFTA, who contended that free trade with Mexico would force lower environmental standards in the United States. The Bush administration had promised not to change any environmental law as a result of the North American Free Trade Agreement. "There is zero interest in making big changes to MMPA," said one congressional aide, and "anything that has the effect of removing the embargo would be very unpopular" (Senzek and Maggs 1991, 3A).

Sensing trouble in Congress regarding the ratification of NAFTA, Secretary of Commerce Robert Mosbacher and others got a bilateral agreement from Mexico to not push the GATT ruling if the United States would try to change MMPA. "Yet word has it that the Bush Administration, which had long been a keen opponent and indifferent imposer of tuna-importing restrictions, first suggested to the Mexicans that they take America's dolphin provisions before GATT" (*Economist* 1991, 31). This was an embarrassing rumor for the Bush administration, which had promised that NAFTA would not reduce U.S. environmental laws. Congress, especially Representative Waxman, lambasted the Bush administration for thinking it could negotiate away the teeth of MMPA (*Economist* 1991). To get fast-track authority, the Bush administration promised no concessions on the environment, but the GATT ruling brings up serious questions of possible "worse case scenarios," according to Waxman (Mathews 1991, A21). The Bush administration also came under attack for its plan to try to amend MMPA at a hearing held by the health and environment subcommittee of the House Energy and Commerce Committee. An administration official admitted that the GATT ruling could be used to challenge other U.S. laws, but that the administration would fight the decision and would consider backing changes in GATT that would prevent similar findings in the future (Maggs 1991).

Most of the embargoed countries came to terms with the law, and international opinion, except Mexico, which appealed to GATT and won, thereby appearing insensitive to environmental concerns just when the Mexican government was trying to convince the United States that it is ready to improve its environmental record. Environmentalists were skeptical about Mexico's increased environmental sensitivity as well as about the administration's promise to defend U.S. environmental laws under NAFTA. Forging a bilateral agreement avoided having to use U.S. veto power in GATT or accepting the GATT ruling (*Journal of Commerce and Commercial* 1991).

The GATT ruling called into question numerous multilateral environmental laws. The agreement is weak on environmental matters, and the original text does not mention the word environment, though it does allow trade measures necessary to protect human, plant, and animal health and to conserve exhaustible natural resources. The GATT panel decided that these words only apply within national borders. A country can use trade measures to protect its own atmosphere, for example, but not the atmosphere outside its borders. This has serious implications for all kinds of environmental laws. Furthermore, GATT requires equal treatment of products regardless of how they are produced—therefore, "tuna is tuna," regardless of how it is caught. Critics of NAFTA and GATT argued that free trade was an indirect way of forcing environmental backsliding that could not be achieved directly. The U.S. administration wanted to avoid a unilateral veto in the GATT council, but letting the ruling stand was out of the question (Mathews 1991).

EC Parliament Supports Environmentalists

In November 1991, U.S. environmentalists celebrated approval by the European Parliament of a report recommending a ban on tuna caught in dolphin-endangering ways, which would stop Italy from importing tuna caught illegally by the Mexican fleet. The recommendation was part of an international battle between free trade advocates and environmentalists. Environmentalists wanted to close Europe—roughly 40 percent of the world's market for canned tuna—to fish caught by purse-seine methods. "It's a victory," said Laura Chapin, spokesperson for the American Humane Society, especially in light of the recent GATT ruling, which environmentalists saw as a setback for MMPA and other environmental laws (Parrish 1991, D1).

David Phillips of EII said that the EII lobbyist at the European Parliament did not expect a vote on the European Commission's report before

December 1991—to become law the report would have to be approved by the twelve members of the Council of Ministers. But one EC commissioner was skeptical that it would become law, since it would be a discriminatory trade barrier, while other EC commissioners said that the vote reflects a "broad political consensus" (Parrish 1991, D1).

Major U.S. tuna packers voluntarily ceased buying tuna caught by dolphin-threatening methods in 1990, and the United States added a formal ban in February 1991, cutting off 50 percent of the world canned-tuna market to the Mexican fleet. Many Europeans were angered at a secondary U.S. ban on Mexican tuna canned in France and Italy. The larger argument pits most countries, including the United States, against international environmentalists over the principle of including environmental standards in international trade agreements (Parrish 1991).

Tuna Laundering, the Secondary Embargo, and More Protests to GATT

In January 1992 federal judge Theldon E. Henderson of San Francisco ordered the U.S. Department of Commerce to ban much of U.S. tuna imports to protect dolphins. The ruling required the United States to impose a secondary embargo on all tuna imports from countries that buy tuna from Mexico, Venezuela, and Vanuatu (Bradsher 1992). The embargo banned imports of yellowfin tuna from any country that imports and exports that tuna. Because yellowfin tuna from Mexico, Venezuela, and Vanuatu had found its way into U.S. markets via "second party" countries, the primary embargo was extended to about twenty other countries (*Facts on File* 1992). The ban covered $266 million worth of imports, which represented about one-half of U.S. tuna imports—only chunk-lite yellowfin tuna was affected, according to Roddy Moscoso, of the Commerce Department (Bradsher 1992).

In February 1992, the U.S. Customs Service via its National Marine Fisheries Service began to enforce the secondary embargo on the countries that processed yellowfin tuna. Those countries included were Canada, Colombia, Costa Rica, Ecuador, France, Great Britain, Indonesia, Italy, Japan, Malaysia, Panama, the Marshall Islands, the Netherlands Antilles, Singapore, South Korea, Spain, Taiwan, Thailand, Trinidad and Tobago, and Venezuela (*New York Times* 1992b). The newly embargoed countries could remove themselves from the embargo by providing "certification and proof" that they had prohibited the import of tuna from the three target countries. Until January 1992, the Commerce Department had required only that importers certify that none of the tuna they bought was originally harvested by one of the target countries and "laundered" through another country (Bradsher 1992).

EC officials protested the U.S. embargo of yellowfin tuna before GATT on February 18, 1992, and said that, along with Thailand and two dozen other countries, they were weighing a new GATT complaint against trade penalties under MMPA (Stier 1992). The secondary embargo ruling affected $4 million to $5 million in tuna exported to the United States from France, Italy, and Britain. The EC pressed for the adoption of the Mexican GATT ruling, even though the Mexicans seemed reluctant to push for the ruling while negotiating NAFTA (Maggs 1992).

The U.S. tuna industry opposed the embargo and had to scramble to find supplies of "unembargoed tuna." "It's an embargo that forces countries to prove they didn't do something. It's guilty until proven innocent," according to Dick Atchison, who represents the American Tunaboat Association (Wastler 1992, 3A). The Bush administration and tuna industry officials appealed the ruling and claimed that the embargo was unfair because it penalized all countries that imported and exported yellowfin tuna, not just those that imported yellowfin tuna from Mexico and Venezuela. Mike McGowan, vice president of Bumble Bee, said, "What the embargo does is prevent us from bringing yellowfin from Thailand into the United States, even though Thailand hasn't bought yellowfin from Mexico and Venezuela in twelve months" (Wastler 1992, 3A).

David Burney, executive director of the U.S. Tuna Foundation said, "We're concerned about the countries where we fish, because obviously the embargo leaves us open for retaliation" (Wastler 1992, 1A). Burney said the tuna industry would lobby for new legislation in Congress that would effectively overturn Judge Henderson's ruling (Bradsher 1992). The Bush administration tried to overturn the embargo, claiming that the action went well beyond the intent of the environmental law on which it was based. This situation further complicated the Bush administration's efforts to bring about GATT and NAFTA and to avoid a domestic and global confrontation over trade and the environment (Maggs 1992).

EII director David Phillips welcomed Judge Henderson's decision as a "complete affirmation of the position we held" (Bradsher 1992, D16). EII also said that the effect of the embargo was overblown by the tuna industry. "We are trying to close all the markets globally for dolphin-death tuna," said Brenda Killian, associate director of the Save the Dolphin Project (Wastler 1992, 3A). Environmentalists asserted that the majority of foreign tuna is skipjack and therefore dolphin-safe and that the absence of yellowfin wouldn't severely affect U.S. canneries. U.S. government officials said that the embargo would affect numerous blended tuna products as well as some fish meal. Environmentalists claimed that only six countries—Colombia, Costa Rica, France, Italy, Panama, and Spain—were guilty of "tuna laundering." Other countries, such as Canada, Ecuador, Great Britain, Indonesia, Malaysia, Marshall Islands, Netherlands Antilles, Singapore, South Korea, Taiwan, Thailand, and Trinidad and Tobago could certify dolphin-safe practices and have the embargo lifted.

According to EII's Phillips, "The Bush Administration is taking intentional steps to create a kind of crisis situation they need to get the embargo overturned and protect their free-trade policy. The whole question involved here between trade and the environment is a potential deal-breaker in the Uruguay round (of GATT)" (Wastler 1992, 3A). "It represents a head-on collision between the Bush Administration's free-trade policy and environmental issues that have been ignored by the Bush Administration," said Craig Merrilees, western director for the Fair Trade Campaign, a group promoting the environment and health laws in U.S. trade agreements (Wastler 1992, 3A).

Embargo Impacts on Mexico and Venezuela

The Mexican and Venezuelan economies were hurt by bad publicity and declining tuna exports. Under MMPA, Venezuela and Mexico could not export tuna products to the United States, because in 1991 their boats had a dolphin kill rate higher than the U.S. fleet average of 1.25. That same year the U.S. fleet killed an average of 2.5 dolphins every time a mile-long purse-seine net was set. Venezuela contended that the United States had set standards for Third World fleets that were impossibly high in order to protect the American fleet when tuna demand was flat (Brooke 1992a, 7).

In response to the secondary embargo, both countries joined IATTC and opened their tuna fleets to inspection by IATTC observers. Mexico pledged $1 million and Venezuela $500,000 for research into dolphin-safe fishing. IATTC reports indicated that the two countries had the largest fleets that set tuna nets in the ETP. The fleet totals for tunaboats in the ETP were Mexico, forty-three; Venezuela, twenty-one; the United States, thirteen; Vanuatu, ten; and Ecuador, eight (Brooke 1992b).

Daniel Covian, a Mexican tuna fisherman, said that he had risked his life for dolphins, even leaping into shark-infested waters to free them from the nets. Now he had to look for a new job because of the "tuna wars" and industry-wide setbacks worsened by a U.S. embargo of Mexican tuna. The conflict threatened hundreds of Mexican fishing and canning jobs (Ellison 1991). "U.S. environmentalists argue that free trade must imply shared values, and that Mexico cannot keep pleading lack of resources as an alibi for wreaking the environment" (Ellison 1991, 5A). Mexican partisans called the embargo a ploy to sabotage Mexico's tuna industry before it challenged U.S. jobs and markets (Ellison 1991).

Years of highly publicized campaigns and boycotts by environmental groups such as EII and Greenpeace had forced most of the U.S. fleet out of the ETP and westward to near New Guinea, where tuna and dolphins stay apart and where the United States has fishing treaties with surrounding islands. Mexico lacked such treaties to gain access to other fishing waters, and Mexican boat owners were not eager to pay the $1 million deemed necessary to equip each boat for such long voyages. Mexico's tuna industry was extremely vulnerable because of the tuna wars and a tuna glut on foreign markets. The price per ton fell more than 30 percent in the first half of 1991 (Ellison 1991).

In Cumana, Venezuela, more and more tunaboats were at dock, and sailors and canners were out of work as a result of the U.S. tuna embargo. Sealed off from the world's largest market since August 1990, Venezuela's tuna fleet had shrunk from 118 boats in 1988 to 34 boats in 1992. As a result of the depressed economy and high rate of unemployment, a crime wave swept the town (Brooke 1992a). According to Laura Rojas, director general of Venezuela's Institute of Foreign Trade, "The U.S. has passed domestic legislation that has jurisdiction outside the U.S. Environmental protection can't be had at the cost of another country" (Brooke 1992a, 7).

Venezuela claimed it had reduced its dolphin kill by 75 percent between 1989 and 1991. According to varying estimates, Venezuelan boats were responsible for between one-fourth and one-third of the 25,000 dolphins killed in 1991 in the ETP. To reduce dolphin kills, Venezuela banned drift nets, placed international observers on all tunaboats over 400 tons, and required all tunaboat captains to take courses on new techniques in minimizing dolphin kills. In May 1992 Venezuela became a full member of IATTC and also pledged $500,000 for research to find better ways to reduce dolphin kills associated with tuna fishing. Venezuelan officials said that the United States had halved its dolphin kill rate over the past decade by shifting fishing from the ETP to the Western Pacific. Claiming to be prisoners of geography, Venezuelan officials said that it was too far for them to fish in the Western Pacific. The ETP primarily includes international fishing waters, while in the Western Pacific tuna captains have to pay royalties and access fees to governments of Polynesian island nations that control local marine resources. According to Eduardo Szeplaki, general coordinator of Fundatropicos, a Venezuelan environmental group, "Under Venezuelan law it is illegal to kill dolphins. Unfortunately, the profit motive leads them to violate the law" (Brooke 1992a, 7).

In February 1992, Venezuela joined members of the EC and twenty-three other nations in urging the United States to abide by a GATT ruling that the unilateral American ban on tuna imports from Mexico and Venezuela was illegal. According to David Phillips of EII, "They are kidding themselves if they think GATT can force the U.S. to abandon laws to protect the global environment. In the 1990s free trade and efforts to protect the environment are on a collision course" (Brooke 1992a, 7). According to Oliver Belisario, a Caracas-based consultant for Venezuela's tuna industry, "Tuna is the debut for a great debate between environmentalists and traders" (Brooke 1992a, 7).

With Venezuelan and Mexican tuna shut out of the United States, Australia, and most of Europe, tuna that landed in Venezuela sold at a steep discount from world prices, e.g., $600 per ton versus $1,000 per ton if dolphin-safe. In 1991 cans labeled dolphin-safe accounted for 95 percent of U.S. sales, and the United States provided half of the world's consumer tuna market. According to John M. Werner, president of the local subsidiary of the H. J. Heinz company in Venezuela, "We don't even can tuna in Venezuela for Venezuelan consumption anymore. Heinz has a worldwide dolphin-safe tuna policy" (Brooke 1992a, 7).

The Second Compromise and the IATTC Accord

In March 1992, the United States, Mexico, and Venezuela reached a preliminary agreement to protect dolphins from tuna fishing nets. GATT officials reported that the three countries had agreed to a five-year moratorium, beginning in 1994, on purse-seine nets (Davis 1992, B10). According to Representative Seade of Mexico, "The main message that should sink in [for environmentalists] is international cooperation," and he hoped that other nations would adopt the accord (Davis 1992, B10).

EII attacked the pact, and congressional aides said the agreements would face a tough time wining approval. David Phillips, director of EII, said the agreements represented a "bad approach," because they would lift trade pressures that had led to sharp declines in dolphin kills. He estimated that 50,000 dolphins were killed through tuna fishing in 1991—about one-half the number killed in 1990. "The current regulatory mechanism is resulting in significant conservation of dolphins," said Phillips (Davis 1992, B10).

Mark Ritchie, a Minneapolis trade analyst who has organized consumer and environmentalist groups against GATT, called the dolphin controversy a "symbol." Even if the United States and Mexico worked out that dispute, he said, environmentalists would still oppose a new GATT agreement because they feared GATT could override some U.S. laws barring imports that don't comply with U.S. environmental standards. GATT officials "have the power to judge and condemn U.S. law from a very limited and undemocratic view," said Ritchie (Davis 1992, B10).

In April 1992 an agreement was negotiated by IATTC, the first major international accord to save dolphins. The nations that fish for tuna in the ETP agreed to cut killing dolphins by 80 percent during the 1990s. From 1980 to 1990, the kill rate was also cut by 80 percent. "The resolution sets into motion a program to reduce dolphin mortality to insignificant levels, to levels approaching zero," according to Dr. James Joseph, director of IATTC (Brooke 1992b, C4). IATTC member countries adopted a system of individual vessel quotas. Under that system, an independent IATTC onboard observer informed the tuna captain when he had reached his vessel quota. Captains who continued fishing were subject to fines or license suspensions (Brooke 1992b). Under the accord, dolphin kills should fall below 5,000 by 1999, down from the 1991 level of 25,000. The accord also called for $4 million to be spent on scientific research projects to help tunaboats catch tuna without killing dolphins. The agreement was reached in La Jolla, California, by representatives of governments that account for 99 percent of the tuna catch in the ETP: Costa Rica, France, Japan, Mexico, Nicaragua, Panama, Spain, the United States, Venezuela, and Vanuatu (Brooke 1992b).

EII argued that the accord was too little, too late. "The reduction is way too little, and the killing of dolphins will continue way too long. In the U.S., consumers, companies, and Congress are saying: eliminate the setting of nets on dolphins," said Phillips of EII. "We do not believe that you can chase down and encircle 1,000 dolphins in a mile-long net and avoid killing them" (Brooke 1992b, C4). According to Richard C. Atchison, executive director of the American Tunaboat Association, the accord is "reasonable, practical, and achievable" (Brooke 1992b, C4).

The Third Compromise

In June 1992, the United States, Mexico, and Venezuela agreed to stop the setting of nets around dolphins and tuna. The Studds bill was introduced in Congress on June 16 and was supported by EII (*New York Times* 1992b). The unlikely alliance of the Bush administration, Congress, environmentalists, and the governments of Mexico and Venezuela forged a tentative agreement to stop the killing of thousands of dolphins caught in purse-seine nets by 1994. After months of negotiations, the bill had bipartisan support and reportedly had already been agreed to by the Mexican and Venezuelan governments. The agreement would end the embargo on Mexico and Venezuela and place a five-year moratorium on purse-seine fishing in the ETP and possibly end purse-seine fishing in the ETP forever (Parrish 1992b).

Environmentalists were optimistic but wary about the new agreement. According to EII's Phillips, the bans and embargoes had prompted the agreement. It was only these "incredible constraints on the market" that forced Mexico and Venezuela, the last countries with big fishing fleets in the ETP, to the bargaining table. "The market for dolphin-unsafe tuna is collapsing. They can't find places to sell the tuna . . . the U.S. won't buy it. England, France and Germany won't buy it. Thailand won't process it, and now very recently some of their last remaining markets in Spain and Italy are collapsing" (Parrish 1992b, 20).

The Studds bill would set a five-year moratorium on purse-seine fishing in the ETP starting in 1994 and would include scientific investigation of alternative methods of tuna fishing and dolphin safety. U.S. and foreign tunaboat owners said that the ban on purse-seine nets in the ETP would cripple their livelihood. "Our vessels and, I believe, the international fleet, would not be able to fish" in the ETP without purse-seine nets, said Richard Atchison, executive director of the San Diego–based American Tunaboat Association. "It's not technically feasible or economically feasible" (Parrish 1992b, A20).

The agreement was a major victory for a small number of environmentalists and for Rep. Gerry E. Studds (D-Mass.), who for years has pushed the Bush administration to enforce the dolphin protection law. Earlier in 1992 EII had rejected a similar compromise that it deemed not strong enough (Maggs 1992). The agreement was politically beneficial for President Bush, as he was being criticized for giving firms more time to meet environmental standards and for opposing many positions at the Rio Earth Summit. The formal agreement with Mexico and Venezuela was expected to be signed after the bill, the International Dolphin Conservation Act, was ratified. In a key concession to environmentalists, Mexico and Venezuela agreed to face stiff penalties if they resumed killing dolphins—a U.S. embargo of all seafood products except shrimp (Maggs 1992).

In the last hours of the congressional session, the U.S. Senate passed the International Dolphin Conservation Act of 1992 (IDCA), which President Bush signed in late October (Parrish 1992a). Although Bush signed the law, "the IDCA will only go into effect if Mexico agrees to comply with its terms, a step which Mexico has so far refused to take" (*Public Citizen* 1993, 9).

Summary

The events illustrated in this chapter underscore the fragmented nature of the U.S. national state. The three branches—executive, legislative, and judiciary—took opposing sides in the tuna-dolphin controversy, as players in the legislative and the judiciary branches resisted the actions of the executive. Additionally, events demonstrated that the environmentalists were able to influence the action of the state despite widespread resistance. Simultaneously, the consequences of the implementation of

MMPA combined with actions of TNCs negatively affected workers in developing countries such as Mexico and Venezuela. The case illustrates how the implementation of proenvironment measures created conditions that demanded reduction of employment and economic opportunities in fishing-dependent regions. Developing countries' nation-states were not able to assist labor through negotiations with the United States, which also indicates the increasing limits of the nation-state in the global system. Finally, the expanding role of supranational organizations, such as GATT and NAFTA, and the limits they impose on various players indicate the embryonic emergence of political entities that operate at the transnational level. Although it is premature to assume that these organizations perform the historical functions of the nation-state, the case illustrates that they provide coordination and mediation in spheres that cannot be managed by national entities.

7

GLOBAL POST-FORDISM: RESTRUCTURING AND THE CONDITIONS OF SOLIDARITY

A number of scholars maintain that we are currently experiencing a period of fundamental change. This change, they argue, is important because it opens possible free spaces that can serve as a forum for the emergence of more just and equitable forms of social relations. O'Connor, for instance, warns us that the current global economic restructuring should not be seen as "merely an objective historical process," but rather as "a subjective historical process—a time when it is not possible to take for granted 'normal' economic, social and other relationships; a time for decision; and a time when what individuals actually do counts for something" (1989, 21). Gordon and his associates argue that "this kind of opportunity for restructuring comes only once in a generation" (1982, 243). Gordon maintains that "we are once again in a position to bargain over institutional transformation" and that the "opportunity for enhanced popular power remains ripe" as "the global economy is up for grabs, and not locked in to some new and immutable order" (1988, 64). Lipietz asserts that we are in a position similar to just before World War II, when the choice was between Nazism and the New Deal. Indeed, we are witness to "a field of tremendous social conflict, whose stake is the negotiation of a new social compromise" (Lipietz 1987a, 51). His principal concern is that "vicious" arrangements will replace "virtuous" arrangements. Piore and Sabel (1984) state that we are at a historical moment in which the foundation of a new social-democratic project designed to codify the new trajectory of capitalist work relations and technology is being established.

An examination of the case of the tuna-dolphin controversy provides us with information on some of the most salient characteristics of global post-Fordism and some of the basic mechanisms through which this new system functions. Additionally, it can guide us in the identification of both emerging contradictions and free spaces for the establishment of more desirable forms of social arrangements. More specifically, the case reveals that the transition to global post-Fordism brings enhanced flexibility at various levels. In particular, the restructuring of the tuna fishing industry first requires enhanced flexibility in terms of the production process, employment, and labor. Second, contrary to hypotheses equating flexibility with disorganization, the tuna-dolphin case indicates that new forms of organization have emerged in the transnational arena. Finally, the case provides information about to the relationship between economic entities (0TNCs), subordinate groups (i.e., environmentalists or workers) and the state. We would like to focus on these three issues, which would lead to a discussion of the broader implications that the emergence of global post-Fordism has in society.

It is our opinion that the case of the tuna-dolphin controversy supports the thesis of "contradictory convergence" (Bonanno 1992, 1993). The case indicates that TNCs find themselves in the contradictory position of having to avoid and, simultaneously, request state support. At the same time, requirements for a more flexible system mandate a significant restructuring of state action, which often entails the elimination and/or reduction of its intervention in a number of socioeconomic spheres. This situation has frequently been defined in terms of the concept "the end of the state" and interpreted as an incapacity of the state to perform its historical roles. The case points out that the national state experiences serious problems when called upon to operate in response to the actions of transnational players. However, the events of the case also demonstrate that the national state still retains critical powers in the socioeconomic arena.

The weakening of the state has important repercussions for groups that oppose TNCs. In the case study, these are represented by workers and environmentalists, who require state intervention to protect their interests. It follows that the presence of a state that can mediate, coordinate, and regulate social relations is requested by both TNCs and subordinate classes, i.e., it is requested by opposing sides of the social spectrum. Due to the crisis of the national state, its organizational and mediative activities must be transposed to the international arena. In this regard the case suggests the emergence of embryonic forms of a transnational state. Moreover, transnational forms of the state and the international arena within which they appear become the contested terrain for both TNCs and subordinate classes. The convergence of interests implies divergent objectives. TNCs' strategies call for regulatory forms that minimize state intervention and, above all, reduce and/or eliminate intervention in support of subordinate classes. Both TNCs and subordinate groups demand a system in which their interests are protected. Additionally, political strength in the international arena is not the same for TNCs and subordinate groups. TNCs' powers exceed those of opposing groups, despite significant resistance such as that carried out by the environmentalists. Resistance to TNCs, the case demonstrates, is also divided as labor opposes the environmental movement's attempts to implement MMPA. Accordingly, new forms of solidarity must be sought in order to unify and strengthen opposition to TNCs. The organization of opposition around class, we conclude, can represent a step toward overcoming the fragmentation entailed in an exclusively pluralistic approach.

Although under global post-Fordism opposition to the actions of TNCs is divided, the case indicates that the global system entails important contradictions, and as such cannot and should not be considered as "totalizing." It is not a totalizing system, as TNCs lack an organizing force at the transnational level that surrogates the functions that the state performs at the domestic level. In this respect, intercorporate competition and transnational mobility become problems that could be addressed through regulation and mediation, which, however, in this context are problematic. Additionally, subordinate groups, such as the environmentalists, successfully oppose TNCs. Although this is not a frequent occurrence, it reflects existing weaknesses of "corporate entities" to pursue their interests at the global level. Although other analyses have taken positions that stress the ability of TNCs to control the global system and therefore have viewed globalization as entailing little room for opposition, this case reveals a much more complex and open reality.

Simultaneously, the case provides indications that local conditions and their constructions cannot be analyzed exclusively from a localistic viewpoint. Rather, this case points out that conditions such as availability of markets, organization of production processes, and local social relations are affected by phenomena occurring in spheres geographically and socially distant from those of the local players. This is not to say that microanalyses do not reveal the richness of relations at the local level and the resources mobilized by local players in the definition of their realities. However, focusing on elements that transcend the local not only adds new dimensions to the understanding of the local situation but also qualifies it by providing meanings to actions such as local diversity and resistance.

FLEXIBILITY

Flexibility of the Production Process[1]

The literature reviewed at the outset of the volume indicates that one of the characteristics of global post-Fordism is the elimination of rigid forms of production. These forms have been gradually replaced by flexible forms that decentralize production while simultaneously organizing it in manners that overcome barriers such as government regulations, high costs of labor, limited availability and high costs of inputs, and limited availability of markets. Examples of flexibility of the production process may be seen in the restructuring of the tuna industry. As the U.S. tuna processors needed guaranteed supplies of tuna during the 1970s, they financed and otherwise integrated with the U.S. fleet. As U.S. production costs increased as compared with those of foreign competitors due to of the implementation of MMPA and the rise of Third World producers, processors divested their interests in the U.S. fleet and relied increasingly on foreign fleets, the international spot market, and low-cost canned tuna imports from Thailand and other Asian countries for their tuna supplies. In effect, as it became increasingly expensive for the tuna TNCs to incur the costs associated with financing and contracting with the U.S. fleet, they cut them off. The new mix of sources gave the firms more flexibility in terms of raw material procurement, labor costs, and governmental regulations.

To reduce production costs, the U.S. industry also closed all but one of its processing plants on the U.S. mainland and transferred production first to Puerto Rico and then to American Samoa, where labor was less expensive, tax incentives were generous, and worker health, welfare, and environmental regulations were more liberal. When Van Camp closed its San Diego plant and Star Kist closed its Terminal Island plant in 1984, about 2,400 jobs were lost. In 1985 Star Kist expanded its Puerto Rico plant to three shifts a day to maintain total processing capacity. The final act of flexible strategy was when the second- and third-largest U.S. tuna firms sold out to Asian firms. These firms' previous strategies had not been adequate to revive acceptable levels of accumulation in the face of stiffening international competition and national regulations.

Flexibility of the production process entails, above all, the ability to compress the spatial and temporal barriers that affected the organization of Fordist forms of production. Global sourcing is a prime example of "spatio-temporal compression" (Harvey 1990). As firms search the globe for low-cost combinations of production sites, processing sites, access to markets, transportation costs, and governmental regulations, their advantageous position with respect to information access provides them with increased leverage over labor and governments. Communications technologies enable TNCs to adjust rapidly to changing political and economic factors worldwide. Knowledge of changing tuna prices on the international spot market influences whether processors offer short-term contracts to tuna fishermen or buy frozen tuna on the spot market.

For example, when the ETP was a profitable and legitimate fishery, the higher costs of labor in California were somewhat offset by the reduced transportation costs and proximity to a large market. As foreign competition grew and MMPA made the ETP more expensive to utilize as a fishery, processors reduced their capacity in California and increased it in Puerto Rico, where labor costs were lower and tax incentives were generous. Even as labor costs increased in Puerto Rico as compared with American Samoa and Asian countries, firms such as Heinz, Bumble Bee, and some Japan-based soga shosha maintained their operations there to take advantage of tax benefits and proximity to the eastern U.S. market. Although they maintained their Puerto Rican canneries, the processors also developed tuna transshipping facilities in the WTP to facilitate more efficient collection and processing of their growing reliance on tuna from that region. According to the U.S. International Trade Commission, from 1982 to July 1991, the number of tuna processing jobs in Puerto Rico dropped from 15,000 to 5,400. This is an average of 1,000 jobs lost per year, which for the most part were concentrated in the city of Ponce, Puerto Rico's major tuna processing location.

At the same time that the tuna TNCs were reorganizing their processing locations, large segments of the U.S. tuna fleet reflagged under Latin

American flags to avoid U.S. observers on their tunaboats and to maintain access to the easy catches in the ETP. Not only did the U.S. fleet reflag, but it also shifted its operations to the WTP and began selling (exporting) its catch to Asian, instead of American, canneries. According to NMFS representatives Herrick and Koplin, this surge in U.S. exports is a classic example of the internationalization of the world tuna trade. In effect, as competition and regulations that hindered accumulation increased, both processors and harvesters "sourced" other arrangements that reduced their costs of production. The Asian firms' actions to buy into the U.S. market are also an excellent example of global sourcing. By having U.S. subsidiaries, Unicord and Mantrust avoid duty costs associated with foreign tuna imports and gain direct access to "fortress USA."

Flexibility also involves the adoption of new technologies that allow an increase in strategic options in the production process. This is the case of the development of technology permitting the production of frozen tuna loins. By importing frozen tuna loins from Asia or Latin America, processors drastically reduced labor costs as well as took advantage of lower tariff structures. Bumble Bee was so enthusiastic about the loining operation that it actually opened a new plant in California. Heinz is increasing its use of loins in Puerto Rico to try to keep its Puerto Rican operations profitable. When Unicord and Mantrust bought Bumble Bee and Van Camp, they instantly gained access to the most modern technologies of tuna production and processing. The labor-intensive processing line is based in the Third World while the (somewhat) more skilled machine tender in the loin plants makes sure the frozen tuna loins get thawed and canned.

Overall, one of the best examples of the use of flexible production strategies is the case of Bumble Bee. Like the other tuna TNCs, Bumble Bee integrated with the U.S. fleet in the 1970s and divested most of its interests in the 1980s. As global competition increased, Bumble Bee opened a processing plant in Manta, Ecuador, to produce loins for its Puerto Rican cannery. Although Van Camp closed its operations in Puerto Rico after Mantrust bought it, Bumble Bee kept its plant open to maintain easy access to eastern U.S. markets. Bumble Bee increased loin imports to Puerto Rico from Ecuador to try to keep processing costs down. It also sources tax incentives in both Puerto Rico and American Samoa, which are duty-free ports. Not only is American Samoa a duty-free port, but foreign vessels can also land their catches there, which they cannot in Puerto Rico. As labor costs increased in Puerto Rico, Bumble Bee shifted its capacity to American Samoa to source the WTP tuna. By processing in American Samoa instead of Puerto Rico, Bumble Bee saves labor costs (minimum wages of $2.82 and $3.53, respectively) as well as lowering the U.S. tuna pack totals that establish the import quotas. More specifically, since the tuna pack from American Samoa is not included in the total U.S. tuna pack, it benefits importers to keep the American Samoa tuna pack high and the rest of the U.S. tuna pack low. Canned imports of tuna packed in water carry a 12 percent rate until import totals reach 20 percent of the previous year's U.S. pack; after that the rate drops to 6 percent.

Bumble Bee also has the flexibility to import tuna loins from its Manta, Ecuador, plant and process them in Puerto Rico where, as a U.S. firm, they have duty-free access to nearby East Coast markets. This takes advantage of lower wage rates in Ecuador and avoids higher wage rates in Puerto Rico. Since Bumble Bee's, and the other tuna TNCs', tax incentives in Puerto Rico decline through the 1990s, at some time Puerto Rico may lose its advantages. In fact, Van Camp and Mitsui recently closed their canneries located there. Bumble Bee can also import loins from its parent, Unicord, the world's largest tuna exporter, and process them at its new San Diego plant designed especially for the loin operation. Tariffs on loins are slight compared to canned tuna. Bumble Bee also imports canned tuna from Unicord and frozen tuna loins to its California plant, which takes advantage of low tariff duties and low-cost labor in Thailand. Additionally, Bumble Bee can either contract with American or foreign vessels for raw materials or buy tuna on the international spot market. The firm was once owned by the American TNC Castle and Cooke, then was sold to its managers, then bought by Pillsbury, which was subsequently bought by Grand Metropolitan of England, which then sold Bumble Bee to Unicord as part of its restructuring in the late 1980s. Bumble Bee is now the second largest U.S. tuna firm but is really a subsidiary of the world's largest tuna exporter, Unicord.

Another example is Safcol of Australia. Safcol has six canneries in Australia and seven in Thailand. It has also previously operated canneries in the Philippines and Indonesia. Because the Australian state manages bluefin tuna catches with a quota system, tuna TNCs such as Safcol have shifted their operations out of harvesting and into financing. The favorable position that the Thai Board of Trade provides export-oriented firms influences Safcol to operate canneries in Thailand. These lease and outsource arrangements give the companies more flexibility to respond to changes in markets, resource availability, and tax/tariff considerations. Whereas before 1980 most tuna consumed in the United States was caught by U.S. tuna boats, processed by American workers, and bought by American consumers, by 1990 tuna consumed in the United States may have been caught by a U.S. vessel thousands of miles from the U.S. mainland, sold to an Asian corporation for processing, and exported into the U.S. in canned form by an American subsidiary of another Asian corporation.

Flexibility and Its Consequences in Terms of Employment and Labor

One of the most significant characteristics of flexible forms of production is the possibility of lowering labor costs and employing labor in ways that were generally not possible during the Fordist era. The transnational restructuring of the tuna fishing industry has been largely carried out in these terms and has had important repercussions, not only in regard to labor costs but also in terms of employment, use of labor, and the overall economic well-being of fishing communities. The introduction of purse-seine net technology in the ETP expanded employment and economic opportunities in the western United States. Later, the passage and contested implementation of MMPA fostered the shift of tuna industry operations to Latin America, with the consequential growth of employment in those regions and the economic decline of tuna fishing operations and communities in the United States. Finally, the embargoes on Latin American producers stimulated a shift of the industry to Asia, curtailing employment and economic growth in Latin American fishing areas. Although through the various phases of the restructuring of the industry various labor pools and socioeconomic regions in the ETP gained and lost employment, the move to Asia signified a serious injury to tuna labor both in the United States and in Latin America. More importantly, the economic decline of the industry both in the United States and Latin America is linked to the implementation of proenvironmental legislation. Paradoxically, for many boat owners, fishermen, canners, and other workers in the industry, the "victory" of the environmental movement is one of the major reasons for their economic problems.

For labor in the United States, TNCs' enhanced flexibility in the tuna industry signified a drastic reduction of employment. During the years of the industry's expansion through the use of purse-seine technology, the high cost of American labor was a problem for tuna companies. The new technology significantly increased labor productivity, while locating production within the United States allowed easy access and proximity to that affluent market. The introduction and implementation of MMPA changed this situation by escalating the costs of production in the United States. However, through various moves, TNCs never separated the issue of controlling labor costs from that of maintaining an entry into the U.S. market. In fact, the decision of the three largest processors to introduce the "dolphin safe" label demonstrated the strategic importance of preserving the U.S. market and of contrasting the advantages of lower production costs with the negative consequences of shrinking markets.

The use of non-U.S. labor in the tuna processing sector is well documented. For example, although Heinz's Star Kist operates the world's largest tuna cannery in Mayaguez, Puerto Rico, U.S. tuna firms are increasingly importing canned tuna to take advantage of lower labor costs in developing countries (Thurston 1990). In 1992, except for Star Kist's Puerto Rican plant, which employs 4,300, the local tuna industry was virtually controlled by Asians: Unicord of Thailand's Bumble Bee; Mantrust of Indonesia's Chicken of the Sea; Mitsubishi of Japan's Caribe Tuna; and Mitsui and Company of Japan's Neptune Packing. For all of these firms, tax benefits under Section 936 of the U.S. Internal Revenue Code are crucial to their remaining in Puerto Rico (Luxner 1990). In Puerto Rico Bumble Bee employs 2,200 workers and processes between 200 and 300 tons of tuna a day, which accounts for more than 50 percent of the Bumble Bee tuna sold on the U.S. mainland.

The move to Asia is a strategy designed to decrease the costs of production and, particularly, the costs of labor. Unicord's low wages at its Thailand factory, where it employs 7,000 people to process raw tuna, allow the Thai company to expand its sales to include affluent markets of the West. "Thai companies, especially in the food-processing business, are aggressively seeking out U.S. companies which control their markets in order to lock up a foothold in fortress Europe and fortress USA," said Graham Catterwell, an analyst at Crosby Securities in Bangkok (Wallace 1992, H3). In these circumstances the usual response of the United States and EC is to apply tariffs to imported goods in order to offset the advantages of cheap labor. To avoid tariffs and to secure footholds in these affluent markets, Asian TNCs have developed another strategy. They buy U.S. or EC companies so that they can be considered domestic companies and, as such, avoid further taxation. Unicord, for example, is Southeast Asia's largest investor in the United States. Before Unicord bought Bumble Bee, it was the world's largest supplier of tuna but was at the mercy of industry middlemen, who bought the fish for resale to major brands. "Now Unicord can be assured of a distribution network in the United States, while Bumble Bee is sure of its supply," said Unicord chairman Kamchorn Sathirakul. "Now we've become a truly integrated, global business" (Wallace 1992, H3).

If the "flexible" strategy to seek less expensive labor is advantageous for TNCs, the same cannot be said for various segments of the U.S. labor force. Boat owners in San Diego have maintained that the big tuna processors' decision to not accept dolphin-unsafe tuna was a tragic decision for them. For twenty years, they had been fighting boat seizures by foreign governments, the closing of American tuna canneries, and foreign fleets slashing prices to capture the U.S. market. However, this last event was the most devastating of all. According to Peter Schmidt, president of Marco Seattle, whose Campbell Industries subsidiary in San Diego is one of the world's leading builders of purse-seine boats, "This could be the last nail on the American tuna boats" (Kraul 1990, D1). Additionally, with the closing of canneries in the continental United States, local tunaboat owners "must now unload their fish at cannery plants in American Samoa and Puerto Rico" (Kraul 1990, D6).

As a result of the consumer boycott of dolphin-unsafe tuna, the three largest U.S. tuna canners turned to Asian suppliers such as the Philippines and Thailand to assure that the tuna they bought had not been caught with purse-seine nets that can kill dolphins. These actions decreased the volume of tuna caught by the U.S. fleet (Thurston 1990). In the year following the big processors' boycott of ETP tuna, the number of U.S. fishing boats in the ETP dropped from thirty to nine (Wallace 1992). After the environmentalist victories of the 1990s, several boats in the U.S. tuna fleet, once the world's largest, went broke and others were sold to foreign interests. Tunaboat captains had to relocate to the Western Pacific and shoulder $1 to $2 million retrofits for larger nets, bigger hydraulics, and new engines (Kronman 1993). Prices paid to tuna fishermen dropped 22 percent to the lowest point in ten years. Within days of the U.S. boycott, the bottom fell out of the tuna market. Yellowfin from the ETP, the best tuna in the world, fell from $1,075 a ton to $835 a ton (Kronman 1993). The shift of sixteen U.S. boats to the Western Pacific and abundant supplies of skipjack and yellowfin tuna increased yields and depressed prices.

Similar considerations can be made for labor in the canning sector. The last six canneries once located in San Diego closed in 1984. By 1987 over 60 percent of the U.S. tuna pack was out of American Samoa. Moreover, by 1993, 40 percent of the U.S.-owned canneries in Puerto Rico had shut down (Kronman 1993). Canneries in Puerto Rico and American Samoa are still open due to the fact that they represent an easy entry to the American market. Once new strategies are developed, even these producing facilities and the employment they generate will likely be eliminated. In fact, by the end of the tuna story, even Puerto Rican and American Samoan labor was getting too expensive (some tax advantages were also running out), and firms such as Van Camp and Mitsui transferred their cannery capacity closer to Asia, where some labor rates (i.e., in Indonesia) are less than $1 per day.

The tuna-dolphin controversy indicates that the enhanced flexibility associated with global post-Fordism signifies more than a simple search for less expensive labor. It involves a much deeper reorganization of capital labor relations, which ultimately goes counter to the enhancement of the overall conditions of workers. First, the relocation of canneries has been characterized by the replacement of male workers with female workers on the processing lines. This is a strategy favored by management since women are paid less than men and are considered a more docile and weaker segment of the labor force. Second, the reorganization of the industry has weakened the ability of labor to mobilize and organize itself. As the tuna firms transferred their processing capacity from California off the mainland, unionized labor in California was systematically replaced with nonunionized labor in Puerto Rico and American Samoa. Third, economic risks have increasingly been shifted to workers. When the U.S. fleet was actually owned by or integrated with U.S. processors, tuna prices were based on long-term contracts. Vessel ownership was directly linked to the firms, either through contract integration or corporate integration. The tuna TNCs provided the financial backing on tuna vessels often costing $10 million. According to specialized reports, when the tuna TNCs divested their fleets to increase their flexibility, the U.S. vessels had to fish on "open tickets" with no guaranteed price for their tuna. Indeed, tuna fishermen often had to negotiate tuna prices with the processors on a per-trip basis, sometimes after they had returned to port. By divesting their fleets, the tuna TNCs transferred the risk associated with vessel operations from themselves to the harvesters.

The negative effects that the transnationalization of the tuna fishing industry has on labor are not limited just to the United States. Latin American countries such as Mexico and Venezuela have experienced significant consequences. As indicated in Chapter 6, Mexico and Venezuela expanded their tuna fishing industry after the introduction of MMPA in the United States. This expansion created new and needed jobs in economically depressed regions. The various embargoes and the overall actions of environmentalists were seen as attempts to attack job opportunities in these areas. They also were interpreted as strategies to enhance U.S. employment to the detriment of Latin American jobs. The various attempts by the governments of Mexico and Venezuela to obtain credibility in terms of dolphin-safe fishing techniques did not succeed in retaining employment in those areas, as TNCs moved their operations to Asia. What remained, however, were strong antienvironmentalist feelings among extended segments of the local population.

In essence, economic opportunities and employment closed for U.S. and Latin American processors as MMPA was increasingly enforced. The industry moved to Asia to source dolphin-safe tuna and low-cost labor, thus marginalizing labor, both for tuna fishermen and tuna processing workers, in the United States (especially Puerto Rico) and in Latin America. Additionally, restructuring of the sector hampered labor's ability to organize and created antienvironmental feelings among workers in North and Latin America.

Coordination in the Global System

The case of the tuna-dolphin controversy offers a series of examples that illustrate the increased flexibility available to transnational economic players. Along with flexibility, however, the case indicates that global post-Fordism is characterized by the emergence of new forms of organization. The issue of whether global post-Fordism is organized or disorganized has been widely debated in recent years. For example, Gordon, Lipietz, Aglietta, and Lash and Urry in various ways and degrees maintain that forms of coordination and organization at the global level are either lacking or are, at best, embryonic. For Gordon, it is "not at all clear that they [capital] have already achieved new global structures of coordination and control" (1988, 25). He challenges the idea of a hypergeographical mobility of capital but acquiesces to the long list of capital's successful assaults on labor. Moreover, he holds that the current restructuring is just the descending slope of a long Kondratieff wave and not the emergence of a new regime: the old is still in decay (disorganized), and the new has not yet emerged.

Lipietz is clear as to his views on the issue of an organized "world regime of accumulation." He claims that (1) there could never be institutional forms regulating world demand; (2) there will never be any supranational authority regulating the money supply; and therefore, (3) a "world regime of accumulation" is only a figure of speech (1987a, 32). Aglietta (1982) adds that international modes of regulation are a much more problematic proposal than nation-bound arrangements. For Lipietz, some form of minimum wage and collective wage are necessary for a successful regime of accumulation. He points out that not even in the EC is there such an agreement.

Lash and Urry argue that disorganization is fostered by the power of globalized capital over nation-states and by the decline of working-class capacities and the associated rise of service-class capacities. The decline of a strong, unified labor movement makes corporatist arrangements much more problematic. The service class is more fragmented and does not vote along class lines. The combination of globalized capital, destabilized nation-states, and working-class decline make corporatist organization impossible. For Lash and Urry, "organization" equals "corporatism." They stress the emergence of international capital as a major hindrance to effective nation-state economic planning but also argue that international capital is not coordinated or organized, but chaotic.

Conversely, Piore and Sabel and David Harvey see new forms of organization and coordination emerging. Harvey indicates that the new regime is characterized by flexible accumulation, which refers to a mix of two basic strategies. First, the extraction of "absolute surplus value" is extended. The latter is achieved through the establishment of longer working hours, family-based sweatshop operations, and wage-relation in precapitalist economies and by moving mass production to low-wage countries and expanding the informalization of the economy. Additionally, flexible accumulation means that corporations pay less to reproduce the work force, with a significant portion of this action being assumed by the state. Second, the extraction of "relative surplus value" is also extended as Fordism penetrates NICs and Third World countries, expanding mass production sites as well as new "flexible specialization" sites. For Harvey, flexible technologies and organizational forms have not "become hegemonic" everywhere, but the current "conjuncture," he argues, "is organized around a mix of highly efficient Fordist production resting on 'artisinal,' paternalistic, or patriarchical (familial) labor relations" (1990, 191).

Piore and Sabel argue that post-Fordism is organized through flexible specialization, which refers to the transformation of labor relations and production systems into a decentralized network of small-to medium-scale cooperatives. Adopting a normative posture, they maintain that a combination of government support, skilled workers using numerically controlled machines, and worker-owned cooperatives producing high-quality specialized products for non–mass markets can successfully integrate with, and even subvert, the dominant and repressive forms of corporate labor organization. Through this system, regional "class compromises" will be established and a democratic transition to post-Fordism achieved.

As far as organization is concerned, the tuna-dolphin controversy indicates that it is erroneous to equate flexibility with disorganization and rigidity with organization. The ability of TNCs to move around the globe creates contradictions at various levels. In this respect, TNCs in the tuna industry found themselves opposed by national governments, judiciary institutions, local communities, social movements, and other social and economic entities. This opposition signifies that TNCs experience difficulties in organizing and coordinating their activities. More specifically, they find themselves in the condition of lacking, at least partially, formal (institutional) and informal networks to assist them in their operations. The Fordist system of tuna fishing could not have prospered with high levels of resistance such as those created by the environmentalists, as it was based on mechanisms that demanded high levels of coordination. As indicated earlier, this action of coordination was, to a large extent, carried out by the state. For example, the U.S. state provided tax incentives and easy access to low-cost labor locations and guaranteed through a number of legislative acts the physical ability of tuna enterprises to operate. The alteration of these conditions of coordination and organization constituted one of the aspects of the transition process from Fordism to global post-Fordism. From this point of view, then, it is possible to conclude that the creation of a flexible system denotes disorganization.[2] However, global post-Fordism also signifies the development of a much expanded system of productive strategic options for TNCs. The activities of Bumble Bee illustrated above are cases in point. Accordingly, it can be argued that not only can flexibility not be equated with disorganization, but global post-Fordism entails the establishment of more sophisticated forms of coordination that are qualitatively different from those of the Fordist era. The ability of TNCs to orchestrate their actions across various countries and labor pools as well as their ability to recombine production processes scattered around the world are examples of the qualitative shift of the new organizational system.

In order to illustrate this point further, it is relevant to recall the ways in which TNCs coordinated their search for less expensive production costs and their desire to maintain an easy entry into the affluent U.S. market. The introduction of MMPA increased production costs, while the move to Latin American countries brought down costs, as less stringent regulations and less expensive labor were employed. The boycotts orchestrated by environmentalists put TNCs in the position of having to choose between continuing their inexpensive production in Latin America or renouncing a significant portion of the U.S. market. Their response was to organize a new, more flexible, and decentralized system in which both inexpensive production processes and easy entry to the U.S. market were maintained. Earlier, this system was built around the use of canneries in tax-safe areas such as Puerto Rico and American Samoa, the eastward shift of fishing areas, and the introduction of new technology such as that for the development of frozen tuna loins. Later it was characterized by the move to Asia as Asian-based TNCs bought U.S.-based tuna companies.

To be sure, the existence of these levels of coordination should not be confused with preordained plans on the part of TNCs to control the sector. Indeed, there are a number of factors that indicate that the coordination of TNCs' activities is problematic and resisted at various levels. First, TNCs in the tuna industry are not unified in the pursuit of particular strategies. Additionally, these strategies vary according to the contingent interests of each company and are often dictated by a complex set of events internal to the enterprise that are guiding the decision-making process. For example, although U.S.-based TNCs such as Ralston Purina decided to leave the sector by selling their subsidiaries to Asian-based companies, TNCs such as Unicord significantly increased their presence in the sector. Second, TNCs compete in the international arena for the control of markets. In this respect, common strategies are possible to the extent that mutually satisfying objectives are sought. Third, TNCs are not immune to attacks coming from various parts of the socio-economic spectrum. The actions of the environmentalists and of the U.S. court are significant examples in this regard. Accordingly, TNCs' original plans of action and their actual behavior are not necessarily coincidental. Finally, and as will be illustrated subsequently in detail, the tuna-dolphin case demonstrated the contested nature of the transnational arena, as neither TNCs nor the environmentalists and the state officialdom are able to fully assert their agenda.

THE STATE IN THE TRANSNATIONAL PHASE OF CAPITALISM

The importance of the state in global post-Fordism finds its roots in the crucial position it occupied in Fordism. In the Fordist system the roles of fostering capital accumulation, providing social legitimation, and coordinating and mediating the actions of various social classes have been carried out primarily by the state. During the period of high Fordism, it appeared to many that the ability of the state to execute these roles was such as to guarantee unlimited stability to the system. The crisis of Fordism highlighted the contradictory nature of the role of the state, while the transition to global post-Fordism emphasized the new limits that it encounters in carrying out its actions. Evidence from the case under consideration reveals the extent and character of these limits as well as the emergence of new forms of that state at the transnational level.

The Weakening of the State at the National Level

From the illustration of the case, it can be concluded that the action of the U.S. state in response to demands from social groups is problematic when lodged in a transnational arena. The bypassing of state action through tunaboat reflagging and industry relocation demonstrated that the ability of the state to perform its historical roles has been weakened. The state is increasingly unable to regulate TNCs' actions (i.e., enforce compliance of MMPA), enhance TNCs' interests (defeat proenvironment groups), and respond to demands stemming from other social groups such as the environmentalists (implementation of MMPA). Also problematic are attempts to extend the state's regulation of economic activities to the international level. The various compromises reached by the U.S., Venezuelan, and Mexican states were designed to respond to the global hypermobility of TNCs (i.e., the move to Asia) and to foster legitimative and accumulative actions at the domestic level (respond to environmentalists' demands in the United States and the loss of employment and economic opportunities in the United States, Venezuela, and Mexico). These territorially limited accords, however, do not match the spatial sphere of action of economic players. TNCs escaped proenvironmental regulations by moving to Asia, where they continue to use the existing purse-seine method and, therefore, can avoid the costly adoption of new, environmentally sound technology. The present situation indicates that TNCs' activities can be regulated in the ETP. Yet they are, at least temporarily, out of reach when operating outside that area. Moreover, despite the existence of multinational accords, TNCs have no immediate interest in reshifting their operations back to the American continent. In essence, through its traditional domestic and international strategies, the U.S. state has very little capacity to control and regulate the activities of TNCs in the tuna sector.

The tuna-dolphin controversy indicates that the transnationalization of the economy implies an economic weakening of the state as well. This tendency is embodied in the propensity of the state to experience renewed fiscal crises. In the multinational phase of capitalism, the state began experiencing fiscal crises as it provided support to multinational corporations, which generated limited taxable resources domestically. More specifically, the economic activities of MNCs were generally located abroad. The metropolitan state thus could not tax the wealth created by these activities. Simultaneously, the nation-state employed domestically generated resources to assist MNCs' actions. This process resulted in increased fiscal pressure on the state and its consequent fiscal crisis. In the transnational phase this tendency continues. The migration of tuna industries to Latin America first and to Asia later indicates an overall loss of taxable activities. This loss is embodied both in the closing of plants and the reflagging and then grounding of fishing boats and in the closing of canneries in the continental United States. The loss of revenues is paralleled by an increased activity on the part of of the state to regulate the sector. The various initiatives by environmentalists, Congress, and U.S. courts are all cases in point.

The closing of canneries in the United States and the industry's move to Latin America and Asia point to the inability of the nation-state to protect the tuna industry and jobs within its territory. The ability of the state to promote the domestic accumulation of capital and to guarantee social legitimation is thus compromised. The capacity of tuna TNCs to source globally for the most convenient factors of production opens up labor pools that are geographically distant from each other. This alteration in the labor market creates a global wage system that hampers attempts by the state to guarantee the continuous existence of higher paying jobs in the country. The availability of cheaper labor in the global market doomed the American-based canneries. Already Puerto Rico and to some extent, American Samoa are too costly in terms of labor wages and rights.

Despite conspicuous evidence indicating a weakening of the state, the same evidence does not support the often discussed hypothesis of the obsolescence and irrelevancy of the nation-state. For instance, although it was detrimental to the U.S. tuna industry, the U.S. nation-state did exercise significant power in enforcing proenvironment legislation. This action was carried out despite opposition from both TNCs and other nation-states such as Mexico. The U.S. state not only enacted legislation to affect U.S. tuna operators but later attempted to extend this legislation to foreign companies that imported tuna to the United States. After the GATT ruling that declared this posture illegal, the Bush administration promised the Salinas administration that, if Mexico would not pursue the GATT ruling, they would work to get the U.S. Congress to amend and weaken MMPA. The Bush administration found very little support in Congress and made no progress in weakening MMPA. In other words, in this case the nation-state was able to exercise its legitimative role against opposing institutional and economic forces.

The weakening of the state vis-à-vis the transnationalization of the economic sphere suggests the problematic character of considering social arrangements as essentially nationally bound. This posture, which has been popularized by the regulationist school, excludes the possibility that social arrangements are grounded in a transnational context. The case of the tuna-dolphin controversy indicates that this possibility exists. Such is the case in terms of labor relations that are defined by the global labor pool and the emergence of global wages. It is also the case in terms of organizational structures and efforts. The organizational activities of TNCs and the emergence of supranational states represent some examples. Finally, it is the very limits that the nation-state encounters in operating within its territory that suggest that social relations cannot be considered, by definition, nationally bound.

The Contested Nature of the State

In a series of recent papers, Luis Llambí (1994; see also Llambí and Gouveia 1994) argues that the transnationalization of the economy has been paralleled by the emergence of three super-states that are hegemonic in global post-Fordism. In his view, there is still a strong identity between the nation-states of the United States, the European Community, and Japan and TNCs that are based in these regions. Therefore, the activities and interests of TNCs are supported by the state, which is unified to further its geopolitical interests. Because of their direct support of TNCs, the intrinsic nature of the super-states is repressive and antipopular, particularly in regard to developing and Third World nations and peoples.

The information available through the tuna-dolphin controversy points to a different set of conclusions. It indicates that the U.S. state is a "relatively open" system where at times subordinate groups "capture" the state apparatus and use it to their ends. In particular, U.S. environmentalists and conservationists mobilized the necessary support for the passage of legislation protecting marine mammals against the interests and opposition of TNCs. Subordinate groups gained access to the legislative apparatus of the U.S. nation-state and, when confronted with opposing interests in the form of the tuna industry and the Department of Commerce, used the legal system to force fractions of the U.S. state to comply with pro-environment laws. The relative openness of the state in this case signifies that the state can have both repressive and progressive elements. The passage and implementation of proenvironment legislation is a case in which the state creates and defends progressive positions against the attacks of economically and socially powerful players.

Additionally, this case illustrates that the state is not a unified system acting in a coordinated manner, as the positions of the three branches of the U.S. state rarely coincided. The executive branch, represented by the Bush and Reagan administrations, opposed the implementation of MMPA, and later the Bush administration was reported to be pleased with the GATT ruling favoring Mexico against U.S. environmental legislation. After the Reagan administration supported a conservative and partial interpretation of the law, U.S. environmentalists demanded that the U.S. state implement a much stronger interpretation of MMPA, i.e., kills reduced to insignificant levels approaching zero. This interpretation was forcefully contested by the U.S. tuna industry and allied departments of the executive branch, which argued that the implementation of this measure would put the U.S. tuna industry out of business. Taking an anti-Bush administration position, the court demanded that the law be applied to foreign tuna vessels and that an embargo be initiated until certification of dolphin-safe tuna could be substantiated. The U.S. Department of Commerce embargoed Mexican and Venezuelan tuna because those nations could not prove they were dolphin safe. Mexico, a country that was already under IMF austerity programs, filed with GATT and accused the United States of trade protectionism; GATT agreed with Mexico. Mexico did not press charges because NAFTA was being negotiated and because it did not wish to be accused of maintaining an environmentally insensitive posture. The U.S. executive branch, which had hoped for a ruling against the United States, promised Mexico to work to get the legislative branch (Congress) to change the law. Congress opposed the Bush administration and did not change the law. In essence, the entire case involves conflicts internal to the U.S. state apparatus in which some of its parts oppose others. From this point of view, the thesis of an identity between the state and any segment of the ruling or subordinate classes is difficult to maintain.

It is possible to equate the position of the state with some of its specific discrete actions. For example, the resolution of the U.S. administration to oppose MMPA can be viewed as proof of the fact that the U.S. state operates in favor of transnational corporations. Hence, it can be argued that there is, indeed, communality of interests between these two players. This, in essence, is the position maintained by Llambí. However, once again the events of the tuna-dolphin controversy do not support this position. First, at the substantive level, discrete actions of the state in favor of TNCs are no more frequent than discrete actions against them. For instance, the position of the U.S. administration against a more radical interpretation of MMPA (i.e., zero dolphin kills versus optimal sustainable populations) and against limitations on foreign imports (i.e., embargoes) has been paralleled by opposing actions carried out at the legislative (i.e., amendments to MMPA) and judicial (i.e., proenvironmental verdicts) levels. It follows that focusing only on one aspect of the entire episode provides a reductionist interpretation of the case at hand. This point leads us to the second objection. At the epistemological level, it is quite problematic to equate one discrete episode with a process. Although an episode is a constitutive part of a process, there are no compelling reasons that can allow us to systematically identify in one episode the essence of the entire process. To be sure, it is possible in some cases that one episode can be emblematic of a process. Yet, there is no evidence that this equation can be applied in every circumstance. Certainly, this is not the case in the tuna-dolphin controversy.

Contradictory Convergence and the Emergence of a Transnational State

Theories of the state discussed in Chapter 3 illustrate a fracture between the sphere of influence of capital (global) and the sphere of influence of the state (national). They maintain that nation-states can control and regulate activities within their territories; however, they cannot do the same for activities that transcend the nation and are global. This dislocation creates limitations on the ability of the nation-state to carry out in society its historical roles of accumulation, legitimation, regulation, and mediation. Such limits, these theories continue, are partially addressed through the emergence of transnational forms of the state, as the sphere of action of the state is extended to match that of economic players. Arguably one of the most discussed hypotheses maintains that forms of the transnational state are emerging, yet are embryonic at best (Friedland 1994; Marsden and Arce 1994). These new transnational states are entities such as the European Community, NAFTA, GATT, and MERCOSUR. Additionally, it is maintained that there is interest on the part of both TNCs and subordinate groups to create an entity that can carry out the role of the state in the transnational arena. This interest, though converging, is contradictory, as it demands the construction of a different kind of state (Bonanno 1993).

Evidence from the tuna-dolphin case indicates it is possible to argue that the emergence of transnational forms of the state could address some of the limits encountered by nation-states in global post-Fordism. Additionally, it is also possible to maintain that emerging forms of the transnational state are currently embryonic and limited. For example, the constrained ability of the U.S. state to control the activities of tuna TNCs is partially countered by attempts to create supranational organizations such as NAFTA, international accords such as those signed between the United States and Latin American countries (i.e., IATTA and the various regional compromises) and the actions of GATT. NAFTA has the potential to attract TNC activities back to North America by linking U.S. capital with low-cost Mexican labor and relaxed environmental regulations. It also can provide a forum for the organization of activities that transcend the national sphere. Additionally, NAFTA can provide the American government with a means to intervene in jurisdictions that were originally outside its sphere of action, as homogenization of regulations could allow the United States to have some control over operations carried out in Mexico. The agreements signed among the various countries involved in tuna fishing in the ETP can represent an international forum for the resolution of controversies such as those that emerged between environmentalists, the tuna industry, and labor. Similar considerations can be made for the role of GATT, which did provide a possible solution to the controversy over the embargo on Latin American tuna.

However, the limited dimension represented by these solutions and the relatively limited power of these organizations are also evident. GATT, NAFTA, and other emerging forms of the transnational state do not have the same powers that the nation-state had in the Fordist period. For instance, despite the more convenient climate established by the passage of NAFTA and the signing of various accords, tuna TNCs have shown no intention of returning from Asia. Additionally, neither NAFTA nor the accords themselves have specific powers outside their geographically limited jurisdictions. Accordingly, TNCs can continue their global sourcing unchecked and outside of the sphere of influence of these new forms of state.

The issue of the role of GATT appears to be different. Despite existing limitations, GATT operates globally. Accordingly, it can match the spheres of action of TNCs and can provide a system of regulation of global activities. However, GATT's exclusive focus on trade and its inability to deal with particular nation-state histories create important limitations. As far as the former is concerned, all socioeconomic dimensions are subordinate to the objective of achieving free trade. As will be discussed later, this situation implies that issues such as the protection of the environment, the protection of labor, and labor regulation are considered subordinate to the establishment of a system of free circulation of commodities. In this context, the legitimation of state action is restricted to fewer options than those available to the historical nation-state. It is possible to hypothesize, then, that mounting legitimation problems could arise. This consideration relates to the fact that GATT is not equipped to deal with the histories of nation-states. The requirements of GATT do not take into account the domestic considerations that motivate nation-states to maintain some form of protectionist policy. The example of the United States during the Bush and Reagan administrations is a case in point. In fact, even though these administrations championed free trade, they never abandoned protectionist postures in a number of significant economic sectors. The case of the maintenance of forms of protectionism though various price support programs in U.S. agriculture is arguably the most relevant example of this type of posture.

The issue of the limited and embryonic nature of emerging forms of the transnational state is accompanied by the argument that there is a convergence of interest from opposing groups in society to create such new forms of state. However, it is also maintained that this convergence of interests is contradictory, as these groups call for a transnational state for different reasons. The basic argument maintains that TNCs are certainly the most powerful players in global post-Fordism; however, their actions generate contradictions. TNCs historically have counted on the support of the state to enhance accumulation of capital and to legitimize these actions to the rest of the population. Furthermore, the state has been instrumental in the control of labor and in generating the legal and social instruments for the availability of labor. In the historical implementation of these actions, the state has been forced to extend concessions to subordinate classes. In this respect, the action of the state in favor of subordinate classes has partially limited its ability to assist corporations in their pursuit of capital accumulation. The contradictory dimension of the relationship between accumulation and legitimation at the domestic level has been partially resolved by TNCs through the bypassing of state action (i.e., global sourcing).

Although bypassing nation-state action partially resolves the contradictions embodied in the relationship between accumulation and legitimation, it opens up new contradictions for TNCs. In fact, the roles of organization and mediation among various fractions of the bourgeoisie that are performed by the state domestically remain unresolved at the international level. TNCs find themselves without an international entity that can reproduce such roles. Furthermore, they encounter the problem of dealing with a nation-state with diminished power to assist them domestically. Evidence from the tuna-dolphin case could be employed to support this argument. For example, in the Fordist period the U.S. state assisted the tuna industry by protecting the American fleet's access to developing countries' coastal fisheries through embargoes and remuneration for lost boats. Moreover, the U.S. state implemented protectionist laws that were designed to discourage imports from foreign countries. In the transnational period, the U.S. state attempted to guarantee further assistance to the local tuna industry; however, resistance from other social actors (i.e., environmentalists) and its limited capacity to operate in the transnational arena hampered its effectiveness. Additionally, in the transnational period some of the problems involving the activities of TNCs in the tuna sector have been resolved, albeit temporarily and partially, through the regulatory action of emerging supranational entities such as GATT.

If it is clear that TNCs could benefit from the emergence of transnational forms of state, the same cannot be said for subordinate groups. As has been demonstrated by the literature reviewed in the previous chapters, the relationship between subordinated classes and the state historically has been a contradictory one. The case of the tuna-dolphin controversy supports this conclusion. On the one hand, the state has been an agent of social control and accordingly, has constrained the actions of subordinate classes seeking the satisfaction of their needs. In the tuna-dolphin controversy, the resisted implementation of MMPA and the various actions of the Reagan and Bush administrations in favor of tuna TNCs are cases in point. On the other hand, the state has allowed the incorporation into society of norms that represent gains for members of subordinate classes. The establishment of social programs, prolabor programs, consumer-oriented programs, and programs in defense of the environment—such as food stamps, unemployment compensation, occupational safety and health (OSHA), EPA, and MMPA—are all cases in point.

In terms of the agricultural and food sector, the role of the state in favor of subordinate classes has been established both in the sphere of the supply of agricultural and food products and in that of demand. As far as the former is concerned, despite important historical and geographical differences, the state has established protectionist systems that have guaranteed minimum levels of income to producers around the world. With respect to the demand for agricultural and food products, legislation has established rules for the protection of consumers, for the improvement of the quality of products, and for the availability of food to needy segments of society. These measures have partially enhanced the overall quality of life for members of subordinate classes. As illustrated earlier, the emergence of a global post-Fordist system implies limits on the action of the state in these areas, hampering the ability of subordinate classes to exercise privileges gained in the past. For instance, the tuna TNCs' tendency to seek less expensive labor abroad jeopardizes the state's effort to maintain adequate wage levels within the national territory. The implications for labor interests are manifold. For domestic labor, the result is a net loss of employment and/or the existence of lower paying jobs; for foreign labor, one implication is the creation of low-wage employment with few workplace protections. For the entire international labor community, there is the constant threat of job elimination through relocation, the decreased possibility of labor mobilization for economic claims, and the creation of a lower "global wage" (McMichael and Myhre 1991). In terms of regional development, there is a decreased possibility of long and sustained economic growth, as local demand and the emergence of external economies are hampered by a system of low wages.

The interest of subordinate classes in maintaining some form of state action in their favor is not restricted to substantive issues such as those discussed above, but is extended to the overall process of participation by these classes in the democratic life of society. The establishment of global post-Fordism has activated processes that jeopardize the availability of "free spaces" for public participation in decision-making processes. In Fordism, the nation-state had maintained free spaces accessible to subordinate classes, which had allowed public participation in the decision-making processes. The state guaranteed the establishment of democratic spaces that historically had been used by subordinated classes to exercise their participatory rights in public life. The possibility of environmentalists using Congress and the court system to further their interests are examples of the existence of such free spaces.[3] In the event that these spaces are closed, the possibility of participation on the part of subordinate classes is severely compromised. Consumer protection, product quality, and the protection of labor and the environment relate to this issue. Indeed, they all represent instances in which the political forum, where the interests of subordinate classes have customarily been articulated, is greatly devalued of its function. The inability of public institutions to enforce measures that directly affect these sectors of socioeconomic life represents a shift of decision-making processes from the public domain to the private sphere. In the public domain the possibility of participation in the decision-making process is available, at least in principle, to all segments of society; however, in the private domain this possibility does not exist. It follows that such a change jeopardizes the continuous existence of effective spaces in which the subordinate classes can exercise their right to participation in the management of society.

The issue of the ability to defend free spaces has been addressed in a number of forums. In the case of the tuna-dolphin controversy, for example, students of international law have focused on the inconsistency between national environmental regulations and international free trade. Trachtman maintains that after the GATT ruled against the United States on the tuna embargo issue, "the United States raised the specter that any rule that struck down the MMPA could also strike down CITES and any similar international rules that are not legislated by GATT" (1992, 148) (CITES is an acronym for the Conference on International Trade in Endangered Species of Wild Flora and Fauna). According to GATT, "U.S. territorial jurisdiction over products cannot be stretched to allow jurisdiction over foreign production processes" (Trachtman 1992, 151). In conclusion, Trachtman argues that "the problem this panel decision underscores is one of legislation: the failure by the international community to legislate comprehensive rules regarding regulatory jurisdiction, and the failure of the international community to legislate rules regarding environmental protection" (1992, 151).

Other authors call for the emergence of some form of transnational state to legislate transnational issues. For example, Holland asserts that in order to remedy the complicated situation surrounding the tuna-dolphin controversy, "an effort on the part of the international community is essential" and "the resolution of this problem should be vested in an international commission" (1991, 277). Although the strengthening of MMPA "illustrates a rare instance whereby unilateral action has prompted individual countries to follow suit . . . , a multinational approach would reduce the potential of tuna fleets circumventing national laws" (Holland 1991, 276, 278). According to Holland, "Essentially, this reduction [in dolphin deaths] indicates that the problem has shifted from a domestic dilemma to one of international importance" whereby "greater efforts on an international level are needed in order to truly protect dolphins" (1991, 279).

McDorman (1991) argues that U.S. fishery import prohibitions raise questions about the consistency of such actions with U.S. obligations under GATT. He states that it can be argued that "U.S. import embargoes promote a conservation-environmental value that is part of an international regime concerning the environment and conservation, which if in conflict with the international trade regime under GATT, is not necessarily subordinate to it" (1991, 507). Furthermore, McDorman maintains that "it cannot be left to the technical rules of treaty interpretation to determine what to do when these regimes collide" (1991, 507). Although laws such as MMPA, CITES, and the South Pacific Drift Net Convention "promote important conservation and environmental values, . . . the embargoes impose values and practices on foreign states that are in conflict with their rights under international law" (1991, 523). Therefore, "there is no compelling reason why the GATT, the international regime concerning trade, should and could not be employed to evaluate the U.S. measures, as they are trade-related measures" (1991, 523). "As the international environmental and conservation regime matures, one can foresee numerous conflicts with the international trade regimes" (1991, 524). McDorman concludes that it is discouraging that the current Uruguay round of multinational trade negotiations does not take into account nontrade considerations and that "the issue of trade and the environment is not prominent" (1991, 525). Finally, according to another student of international law:

> The international trade regime is in one sense inherently incompatible with environmental protection laws. The objective of international trade law, as embodied in GATT, is the free flow of goods and services with a minimum of governmental interference. In contrast,

> environmental conservation advocates seek increased governmental regulation to protect the environment. This conflict is becoming increasingly common as environmental consciousness grows. Many environmental organizations see the GATT and its free trade principles as a threat to national environmental laws and international environmental accords, while free trade advocates complain of "environmental imperialism." The Mexican tuna embargo is a prototypical dispute. (Black 1992, 125)

From the above, it is evident that the interests of subordinate classes for the creation of a transnational state are grounded in their interest in protecting the gains they have obtained in society. TNCs' interests, on the other hand, are based on the support they can receive from the state to enhance capital accumulation. The contradiction between the current trade priorities upon which GATT is organized and the maintenance of national proenvironment laws underscores the fact that social groups can consciously affect the evolution of global post-Fordism. More specifically, the creation of supranational forms of state such as GATT and NAFTA is a social process that is constructed by the interaction of relevant social forces. However, this contradiction has also shown that TNCs and subordinate classes do not have the same historical possibilities of affecting the creation of a transnational state. Through the years, TNCs have developed transnational networks that have enabled them to partially reproduce the process of coordination and organization provided by the state domestically.

This case study points out that the organizational and coordinating ability of transnational corporations is significantly developed and that their capacity to influence political decisions is greater than ever. Indeed, GATT's exclusive focus on trade is a construction that ultimately favors TNCs' interests in eliminating state barriers, such as those imposed domestically by the environmentalists, and in increasing overall flexibility of operations. It can, therefore, be concluded that the current functioning of GATT, vis-à-vis the tuna-dolphin case, fosters the interests of TNCs against those of environmentalists. In essence, despite common interests in building a transnational state, TNCs and subordinate classes enter this historical phase in a position of conflict. The contradictory dimension of the convergence of interests involves contrasting goals that are at the core of the agendas of both groups.[4] As far as subordinate classes are concerned, they have established a presence in some of the emerging transnational political institutions, such as the European Community. Despite the fact that the presence of subordinate classes is occasionally quite significant politically, they generally retain the status of political minority that they have at the domestic level.

THE TERRAIN OF CHANGE: NEW PLAYERS, REGULATION, SOLIDARITY, AND CLASS

The emergence of global post-Fordism also entails changes in the principal social players. TNCs operate in a transnational arena and are qualitatively different from the multinational corporations of the Fordist era. As illustrated earlier, their power in regard to nation-states has increased significantly as well as their ability to counter actions of opposing groups. Opposition to TNCs comes from a variety of players. The tuna case demonstrates the power of environmentalists, who were able to impose a stronger implementation of MMPA and limit the action of TNCs. Although the tuna industry and supporting departments of the U.S. state fought against MMPA for twenty years, through persistent use of the U.S. court system, consumer boycotts, and coalition building, the environmentalists did stop the use of purse-seine nets in the ETP.

The increase in the political power of environmentalists and other new social movements (NSMs) has led some writers to argue that globalized players' counter to global post-Fordism is not labor but the "new social movements" (Buttel 1992; Gorz 1982; Lash and Urry 1987; Lipietz 1992). The core of the new social movements includes the environmental movement, the feminist movement, the antiimperialism movement, the antiracism movement, the antinuclear movement, and the peace movement. According to Buttel, "most scholars have defined these movements in terms of a rejection of the institutional practices of the modern state and economy" (1992, 10). NSMs emerged during and in response to the transition to post-Fordism "as a consequence of the failure of social democratic parties to undergo a political process of renewal" in the face of the decline of social-democratic, or Fordist, societies (Scott 1990, 25). In addition, NSMs are seen by some authors as a powerful counter to the rise of neoconservatism characteristic of the transition to post-Fordism (Buttel 1992; Giddens 1987; Scott 1990). As it will elaborated later, the case study indicates that NSMs have serious limitations in representing the interests of other subordinate groups. Indeed, in this case the "victory" of the environmentalists alienated labor support and damaged the socio-economic well-being of fishing communities in developed and developing countries. Simultaneously, environmentalists lacked the vision for the inclusion of other subordinate groups in their political strategy. It follows that proposals stressing the key emancipatory role to be played by NSMs (e.g., Buttel 1994, 24–30) find little support from the evidence pertaining to this case.

The growth of NSMs has been accompanied by an important weakening of organized labor. This situation has led several authors to argue that organized labor is a casualty of the transition to global post-Fordism (Gordon et al. 1982; Lash and Urry 1987; Offe 1985). Indeed, for a number of writers, global post-Fordism is a direct and successful attack on labor and labor's protective welfare-state (Giddens 1987; Gorz 1982; Lash and Urry 1987; Offe 1985). The tuna-dolphin case supports this conclusion, as labor has been defeated in both advanced and developing countries. It was defeated in the United States, as canneries closed first in the continental United States and later in Puerto Rico. Additionally, the restructuring of the tunaboat industry meant that boat operators released their crews and abandoned the sector. Labor has also been defeated in developing countries, such as Venezuela and Mexico. In Latin America the crisis of the U.S. industry represented the creation of new jobs in the 1980s. The move to Asia halted this process and initiated a devastating crisis for fishing communities along the ETP. Moreover, labor has been split, as workers and fishing communities have been pitted one against the other. It was hoped that through this competition some of them could attract TNC investments and translate them into jobs. However, TNCs' actions have certainly hampered these hopes for the American continent. Traditional forms of aggregation and state protection of workers' rights and status have been made increasingly ineffective by global restructuring. The loss of jobs and global sourcing by TNCs have greatly diminished the ability of labor organizations to mobilize workers. And even in cases in which mobilization has occurred, the labor movement has not accomplished most of its goals. The mobilization of Latin American workers to save employment in local canneries and on fishing boats and the mobilization of workers in California to save U.S. tuna industry–related jobs are cases in point.

We have argued earlier that the transnational arena is contested, as neither TNCs nor the environmentalists and the state officialdom have been able to fully assert their agendas. The limits of the regulatory capacity of the nation-state have been shown, along with the difficulties that exist in the development of larger-than-national forms of regulation. Following the tuna-dolphin case, it can be argued that the regulatory situation at the transnational level is extremely unsettled and characterized by a combination of old forms of regulation paralleled by emerging new ones. The former refers to the various nation-states' attempts to continue their mediative and organizational roles both domestically and internationally. New forms of regulation are embodied in the increasingly important role performed by transnational organizations, which in this particular case refers to GATT and IATTC. The unsettled character of this situation is supported by the inability of these institutions to maintain levels of control that encompass the sphere of action of TNCs and that address the demands from other social players. Overall, the case study demonstrates that issues concerning the protection of the environment and labor cannot be addressed unless some form of control of TNCs' actions and the regulation of the overall system are carried out. More specifically, this case points out the validity of the assumption indicating that unrestricted development of capitalism creates unbearable consequences for society.[5]

If the scenario is correct, then the task of finding solutions to control and regulate the current trends of global post-Fordism assumes central importance. Gordon et al. (1982) argue that the emergence of a new political coalition is necessary to resolve the crisis and to find new ways to control capitalism. These authors saw the rise of the New Right under President Ronald Reagan as a possible precursor to the new social structure of accumulation (SSA). The SSA, like the European regulationists' concept of the mode of regulation, is the set of institutional matrices that coordinate accumulation, that provide legitimating and mediating functions. For Gordon et al., the Reagan administration's strategy to revive accumulation by "rolling back the gains" achieved by labor under the welfare-state provided both the philosophy and the policies that could underpin a new SSA. They worried that such a New Right SSA would ensure continued "widespread hardship and risk social catastrophe" unless the Left successfully advanced its own political agenda creating a new SSA (1982, 243). Gordon argues that we must resist characterizing the power of TNCs as unassailable, because it handcuffs us "with misplaced and potentially crippling perceptions about the new global order" (1988, 26).

Piore and Sabel see two possible paths to resolve the contradictions of transnational capital. One route requires the construction of a new world economic order called "international Keynesianism," where mass production continues to dominate in the following contexts: an amalgamation of markets into regional trade blocks, coordinated fiscal and monetary programs, a strengthened role for IMF and GATT that would provide Bretton Woods–like stability, and a new set of understandings between the developed and under-developed nations that would balance global supply and demand and apportion economic growth around the world (1984, 252–57).

The other route is represented by flexible specialization, or "yeoman democracy," where the institutions of management, labor, and government are reinvented to support the infusion of technology into the workplace so as to promote more flexibility in production and greater skills in the work force. New social relationships would arise, reducing the distance between worker and boss and encouraging the introduction of technology that enhances the creativity of the workers. These new relationships would be regulated by a neo-Keynesian state similar to the New Deal (1984, 258–76).

Piore and Sabel conclude that it is possible that "flexible specialization and mass production could be combined in a unified international economy . . . [where] the old mass-production industries might migrate to the underdeveloped world, leaving behind in the industrial world the high-tech industries" (1984, 279). Although such a hybrid system would not last forever, "it would create a universal interest in two basic goals: worldwide prosperity and a transnational welfare state" (1984, 280).

Lipietz (1987a) also sees two possible paths facing global post-Fordism. One strategy focuses on a productivist model based on neo-Fordism, and another is grounded in the mass extension of nonwage-earning activity, such as leisure, creativity, and intellectual enrichment. Lipietz presents the futurologist Herman Kahn's prediction of the rise of a new Kondratieff cycle based on the automation/electronics revolution centered around Japan and its satellite NICs as an example of the productivist model. He argues that this "virtuous configuration" between Japan and the Asiatic NICs is only maintained "by pumping capital out of the rest of the world" and that, therefore, "no virtuous global cycle has yet emerged" (1987a, 50).

Although the neo-Fordist robotized workshop can rapidly adapt itself to changing consumer tastes by reconfiguring production runs into short series, the particular form that the flexible automated systems take has followed two different strategies with critical implications for workers. In the United States and European states such as France and Great Britain firms have used neo-Fordist automation to further separate the intellectual and manual tasks of production—an extension of Taylorism and the intensification of labor control. This strategy is designed to generate profits by targeting massive investments based on short-run production series destined for a segment of well-off consumers. For Lipietz, by focusing on a relatively narrow group of elite consumers who can afford the electronic products, this strategy ignores the requirements of mass consumption, fosters chronic unemployment, and eventually results in a society divided into three classes: elites who can afford the new high-tech products, a small, privileged stratum of labor that makes the products, and a growing mass of marginalized workers (1987a, 52). The other path chosen by Japanese firms and European firms in Germany and the Third Italy has been characterized by reincorporating workers' intellectual activities into the automated production process. This path apportions productivity gains to workers in the form of reduced work time and increased mass access to consumptive goods. This path would be less competitive because of increased wage costs.

For Lipietz, the pursuit of both paths could result in two possible outcomes. First, an unending struggle for economic dominance could lead to a extended trade war between the firms and nations committed to each strategy. Second, an international accord could be reached that would permit each strategy to proceed in its own relatively delinked system. Lipietz provides the example of IMF's policy of austerity programs instead of debt cancellation or freezing, as its strategy for resolving the Third World debt crisis as an indication of the more likely path—the repressive one where middle classes are pushed "back into poverty, the workers towards wretchedness, and the poor towards death" (1987a, 52).

The tuna-dolphin case allows us to provide additional possible scenarios to counter the undesirable consequences of the development of global post-Fordism. The difficulties that the political sphere has in controlling the economic sphere is a crucial starting point. In this context, fully developed forms of control and regulation are possible only in the event that the two spheres match each other. In other words, the political sphere should have the same range of action as the economic sphere. There are two general categories of alternatives that can be derived from these assumptions. The first is to develop some form of protectionism at the national and or supranational (i.e., NAFTA, EC) level. Protectionism can be advocated to limit the transnational mobility of capital and labor. In recent years, protectionist strategies have been called for for items such as the adoption of increased import tariffs, tougher controls for commodities and labor at the border, and various incentives to enhance domestic production and consumption of domestic products (i.e., buy American). However, in light of the ability of tuna TNCs to shift operations to more "friendly territory," important objections to this strategy can be raised. The limits of the protectionist strategy can be synthesized into two objections. The first is that protectionism counters global players' post-Fordist strategy of flexible patterns of accumulation, which are keys in overcoming the crisis of Fordism and expanding the avenues for the growth of capital. Accordingly, protectionist strategies would involve hampering the functioning of a system that is increasingly interrelated globally and that finds its ultimate strength in its global character. The second objection refers to the nation-states' decreasing capacity to enforce local legislation. In this case, as documented above, there is no reason to justify the conclusion that protectionism would enhance the state's capacity to effectively enforce its regulations in the international arena. Indeed, the capacity of the state to protect the interests of subordinate classes and to simultaneously assist the actions of TNCs cannot be guaranteed with political measures alone. The structural and systemic conditions that contributed to the development of global post-Fordism must be addressed as well. Protectionist strategies do not address these conditions.

The other general alternative is to be found in international accords aimed at regulation. This alternative can address the proposal that the unity of the polity and the economy must be reconstructed. It would involve the creation of transnational polity forms that surrogate the functioning of the nation-state at the transnational level, such as IATTC. In essence, this alternative would involve the creation of international alliances or organizations that would control TNCs' capacities to exercise flexible strategies to their advantage. In order to do this, one of the major aspects to be overcome would be the fragmentation emerging in production and cultural spheres (Mingione 1991). As indicated earlier, progressive movements such as environmentalist and prolabor organizations, as well as local communities, are pitted one against the other. Accordingly, the communalities shared by these movements and communities, both at the economic and solidarity levels,[6] are weakened by the emphasis on locality and particularity of interest. This situation matures in a context in which the interests of transnational capital are not fully criticized. The case shows that, although the U.S. national controversy was centered on the banning of established fishing techniques, it ignored domestic labor issues. Alternatively, the international controversy barely touched the issue of the industry restructuring around new technologies, focusing instead on the localized impacts of labor dislocation. The international discourse centers on local advantages and gains of local groups without questioning the tuna industry's insistence on purse-seine technologies. Gains are framed in a taken-for-granted discourse that addresses immediate concerns but never embraces the more probing issue of long-term social arrangements. In other words, the objective of profit generation is ultimately maintained along with the "alternative" goals of constructing a sound environment and developing poor world regions.

The crisis of labor and the emergence of new social movements constitute two of the most important features of global post-Fordism. However, the events of the case indicate that the environmentalists and labor oppose each other. Indeed, various segments of labor, both in the United States and abroad, identify environmentalists' actions as the source of declining employment and business opportunities. They view environmentalists' requests for implementation of MMPA and their various boycotts as direct attacks on labor and on the survival of entrepreneurial activities, such as the existence of tuna fishing boats and canneries. The sign held by a Venezuelan woman claiming that U.S. environmentalists starve Venezuelan children to save Flipper is emblematic of the situation. Simultaneously, the environmental movement has paid little attention to the economic consequences that its actions have had on fishing communities and workers. Environmentalists viewed economically related issues as ultimately secondary in comparison with the primary task of protecting the environment. Both movements, however, reacted to the fact that TNCs took full advantage of global sourcing and that they ultimately avoided environmental restrictions by shifting their operations to Asia. The implementation of MMPA is certainly a victory for the environmentalists; however, the fact that TNCs have not abandoned the existing fishing technology and that they moved operations from the areas where proenvironment legislation was implemented are signs of the limits of the environmentalists' "victory." In essence, the case indicates that the process of controlling the actions of TNCs requires a coalition of broader forces than those represented individually by the environmental and labor movements.

The labor-environmental split calls into question the existence of conditions for a broader solidarity among forces that oppose TNCs. The case demonstrates that Latin American and U.S. workers do not share and comprehend the environmentalists' emphasis on the protection of the environment over the creation of jobs, over the creation of strategies for the economic stability of fishing communities, and over the difficulties of developing countries in adapting to environmental standards created in the First World. The environmentalists' posture reflects broader discourses that have emerged in global post-Fordism. Particularly, it reflects the increasing shift toward the noneconomic in the name of more culturally centered discourses. The failure of social and political projects that identified the economy as the central element for the reorganization of society created the conditions for an emphasis on the cultural (Harvey 1990). Postmodern discourses that emerged forcefully in the 1980s capture this shift (Baudrillard 1983). In these discourses emancipation has often been equated with the efforts of various minorities (e.g., environmentalists, women, ethnic groups, and so on) to defend and eventually extend their rights in an increasingly oppressive society.

Pluralistic solutions have been proposed in the name of a multicultural worldview and have been considered the most advanced forms of democracy and the most powerful alternatives to corporate domination. Despite these conclusions, pluralism remains a compromise of difficult realization. This situation is due primarily to the almost insurmountable difficulties in reconciling demands for the protection of diverging rights. Indeed, the debate on pluralism has been characterized by the emergence of increasing theoretical and political divides. The ensemble of interests and values emerges as too complex and divergent to foster coherent proposals of action. In an effort to avoid conflict among the various groups, attempts to develop such proposals have been created in terms that are too abstract to provide the necessary guidance (e.g., Habermas 1987, Laclau and Mouffe 1984). An emancipatory coalition based on the protection of various groups' rights requires specific aggregative solutions.

The latter have been grounded in key concepts such as those of tolerance and difference. This case demonstrates, however, that tolerance and difference have failed to forge a broader aggregative movement against TNCs, as the two historically progressive movements involved in the case positioned themselves against each other.[7]

The difficulties of expressing anti-TNC solidarity on multicultural grounds open up the possibility of reconsidering class as an aggregative element. Indeed, in the case of the tuna-dolphin controversy the harsher socioeconomic consequences materialize in a context in which environmental and labor issues intersect with class. Certainly, the concept of class in today's sociological debate has lost most of the power that it had in the past. In classic sociology, for instance, the importance of class for the creation and maintenance of stability and economic growth in society was central. Regardless of their theoretical postures, classic social thinkers were extremely aware of the fact that significant class division would jeopardize the existence of fundamental democratic institutions. Moreover, they were aware that sharp social division would often entail the use of repressive force, which would create limits to free circulation of labor and commodities. Ultimately, they believed, repressive political systems put limits on the growth of capitalism. Post–World War II sociological theory further emphasized the importance of diminishing class polarity. In these years, the concepts of modernization and development were all grounded on the assumptions of diminished polarity and of an expanding middle class. Even postindustrial theorists argued that the end of sharp class differences separated classic capitalism from the more affluent and stable postindustrial phase. Conversely, in many circles today the concept of class has been set aside for more fashionable theoretical constructs. Class is limited to references to the middle class or is simply dismissed in the name of a free market system that provides opportunity for everyone. Moreover, established systems to correct social inequality are increasingly viewed as wasteful and even antidemocratic. However, this theoretical preference clashes with empirical evidence that indicates increasing poverty, marginalization of ethnic, gender, and other minorities, and the crystallization of an economy based on chronic underemployment and unemployment.[8]

Certainly, a concept of class that simply classifies individuals or aggregates into categories will be of little use. Moreover, ahistorical visions of class inspired by new renderings of vulgar economicism should also be dismissed. The concept of class should, instead, be interpreted in terms of the conditions for the creation and reproduction of "modes of life" (Marx 1947). In this sense, class refers to the ability of groups to reproduce and maintain their positions in society and to the conditions that prevent the realization of such circumstances. Socioeconomic trends of the past decade indicate that the possibility of reproducing their modes of life has diminished for growing segments of the population, both in developed and developing regions. This condition cuts across various sites of exploitation and domination to indicate that the most serious conditions are found at the intersection of class with gender, race, ethnicity, environmental degradation, and other relevant characteristics of life. The reintroduction of a discourse in which class is a primary, but certainly not exclusive, element for aggregation is an attempt to find a common ground that multicultural approaches and their emphasis on cultural fragmentation do not provide. As indicated by classic thinkers such as Marx, Durkheim, and Weber, solidarity among social groups must be created, reinforced, and maintained to defeat the fragmenting and alienating tendencies of modern society. In global post-Fordism the acceleration of centrifugal tendencies is overwhelming. The market cannot regulate itself and social players call for regulatory entities. A common language of solidarity must be found, and class might represent a solution, as uncertainties for subordinate groups mount under global post-Fordism.

NOTES

Chapter 2. The Debate on the Transition from Fordism to Global Post-Fordism

1. Antonio Gramsci (1971, 277–318) first used the concept in the late 1920s primarily to avoid fascist censorship. He implied a distinctly new form of very highly rationalized capitalism, combining mass production, planning, and bureaucracy. The term was later reintroduced by the "regulationist" school (e.g., Aglietta 1979 and Lipietz 1987). Yet it has been used in different ways by many interdisciplinary social theorists. Our usage in this chapter and in the book does not follow that of the regulationists.

2. The "state" should not be understood in terms of a singleminded and unitary actor. Rather, it has multiple levels and forms, many semiautonomous agents and offices, and its policies are a contested terrain influenced by a complex of competing interests.

3. The dependent position of Third World countries is central in the maintenance of Fordist growth in advanced societies. In this respect, it is misleading to interpret Fordism in exclusively "national" terms. Though the characteristics of Fordism illustrated in these pages refer to advanced nations of the West, Fordism should be viewed in the context of the exploitative relationships between developed and developing countries in a world system.

4. Parsons (1971, 107–8) does not speak explicitly of Fordism but discusses Henry Ford's essential contribution to the creation of the "lead society." The huge literature on international "development" (e.g., Hoselitz 1960; Rostow 1960; Allen 1974; Inkeles and Smith 1974) inspired aid programs worldwide based on these principles. The key for the success of developmental projects in the first three postwar decades seemed to be the imitation of U.S. strategies for socioeconomic expansion. The social psychological predispositions also had to be imported to pave the way for organizational and material transformation. These

social and political ("modernization") theories of development played an active role in the formation of American policy toward the less developed world during this period.

5. Bell (1976) also spoke of the disintegration of the Protestant ethic and of the workaday values that constrained the disintegrative and hedonistic aspects of capitalism. He contended that the postmodern cultural environment of the 1970s accelerated and unleashed hedonistic tendencies, already present in earlier capitalist consumption, which undermine the basis of the capitalist labor and production system. This argument points back to Marx's ([1848] 1947, 12) argument that capitalism causes "everything solid to melt into air." At the same time, Bell departs from Marx, contending that capitalism, in order to stay on course, requires disciplinary, cultural resources to perpetuate the work and organizational networks necessary for the maintenance of capitalist production. Much to his dismay, Bell's argument has been viewed as a major contribution to the neoconservative movement of the 1970s.

6. In the present analysis the terms "global post-Fordism" and "post-Fordism" refer to the same historical situations. Accordingly, they are synonymous.

7. But this trajectory of change is an ongoing process involving disparate experimental strategies in many different locations. Because current conditions contain too many uncertainties, it is premature to argue that a distinctly new stage of capitalism has been consolidated or that all the important elements of the Fordist regime are in permanent eclipse.

8. See Harvey's (1990, 284–307) many examples on these matters—e.g., satellite communications systems, cheaper types of air freight and bulk overland and water transport, the dematerialization of money, and new information and video commodities.

9. For example, poultry producers avert dealing with very high environmental costs by seeking sites where environmental legislation is less stringent. Hence, they selected Arkansas over nearby Kansas, and should regulation be increased in Arkansas they might move to yet another state or outside the nation (Giardina and Bates 1991). It is also easy to see how decentralized operations facilitate this strategy.

10. Note, however, that companies also enjoy savings from such arrangements. For example, homeworkers often use their own equipment and bear the costs of its repair as well as much broader, basic operational costs (i.e., heating and cooling, wear and tear, insurance, and other factors related to the home). Even the tax deductions constitute a subsidy for the firm.

11. Corporations are mobile as well. In its annual evaluation of American companies, *Forbes Magazine* lists "Japanese" Honda as the fastest-growing U.S. company for 1991. It is considered American, because the company employs 14,000 American workers and half of its sales are in the United States (Flint 1992). At the same time, many models of traditional "American" companies are produced abroad either independently or through joint ventures with "foreign" corporations. For example, the Ford Crown Victoria and the Mercury Tracer are

made in Mexico, the Ford Festiva is made in Korea, and the Ford Probe is made by Mazda in Japan; General Motors owns a portion of Isuzu, and the "American" Saturn plant is a joint venture with Toyota. Another example can be drawn from the food sector. Tyson uses feed produced in China to raise chickens in the United States, which are deboned in Mexico and then sold throughout the affluent nations of the West (Constance and Heffernan 1991). Advocates of the "buy American" movement cannot easily distinguish which products are truly indigenous. Even some companies that produce, process, and sell goods entirely in the United States employ exclusively or mainly low-wage labor imported from Third World countries (e.g., Gouveia 1992).

12. Postmodernists offer new critical perspectives about liberal welfare democracy as a system of social and cultural control. They have also contributed to opening new cultural spaces for differentiated life-styles, values, and expression. On the other hand, many postmodern writers too readily celebrate the current fragmentation as an unleashing of difference and plurality, which they claim follows from the collapse of totalizing Enlightenment ideologies (about emancipation and democracy). They tend to treat modern political solidarities and the forms of state mediation of capitalism to be repressive forces that merely facilitate domination, conformity, and the overall repression of difference. Reviving facsimiles of Fordist arguments about postindustrialism and the end of ideology, some postmodern writers have declared an end to politics and to the entire Enlightenment tradition. The atomistic, individualistic, and crudely commercial side of some postmodern thinking reflects the alienated and exploitative "liberty" and "diversity" of class-polarized consumer culture.

13. Although we find many elements of Jameson's and Harvey's broader arguments extremely useful, we do not see postmodernism as merely the "logic of late capitalism" or simply a manifestation of accelerated "turnover time." Postmodern culture, although part of the global post-Fordist phase of capitalism, is not entirely economically "determined" and has also been shaped by noneconomic, relatively autonomous cultural developments.

14. See Bartlett and Steele 1992, Bradbury 1986, Danziger et al. 1989, Ehrenreich 1989, Mishel and Bernstein 1993, Newman 1988, and Strobel 1993 for copious data on the growing class polarization and declining middle class.

15. The Parisian regulationist school is not the only regulationist approach present in the European debate. Five research groups—Grenoblois, PCF-CME, the Amsterdam school, the West German school, and the Nordic school—deserve at least a brief discussion.

The Grenoblois group has been researching regulation in capitalist societies since the mid-1970s (GRREC 1983). This groups adopts two primary reference points: a critique of the theory of general economic equilibrium as an adequate basis for understanding capitalist dynamics and a periodization of capitalism into three stages, each with its own particular mode of regulation. In arguing against a general economic equilibrium theory with its tendency to operate outside real time and space, the Grenoblois assert the need for social procedures of regulation that secure expanded reproduction of capital for limited periods within a given economic space. The three periods put forth are competitive or liberal capitalism, simple monopoly capitalism, and state monopoly capitalism. The social procedures of regulation characteristic of each period must maintain an adequate rate of profit for all sectors of capital in the face of capitalist competition and secure a tolerable balance between the structures of production and consumption in the face of class struggle. Because each stage needs its own mode of regulation, crisis in the social procedures that compromise a particular mode of regulation will trigger struggles to find the next viable mode of regulation (Jessop 1990).

The PCF-CME approach is inspired by Paul Boccara and the French Communist Party (PCF) (Boccara 1983, 1985; Drugman 1984; di Ruzza 1988). This school developed in the 1960s and presents a new view of state monopoly theory based on a law of "overaccumulation-devalorization" and its impact on relations between the state and private monopolies. According to PCF-CME, overaccumulation is inherent in the tendency for the rate of profit to fall, which occurs because the prevailing relations of production block the development of new productive forces. In the short run, overaccumulation is countered by capital's reorganization of the labor process; in the long run it is eliminated by the devalorization of a part of the total social capital. In the current period of state monopoly capitalism, devalorization is the state's responsibility and is secured through public finance or subsidies for monopoly investment, nationalization of key infrastructural sectors to provide inputs below costs of production at the expense of higher charges to nonmonopoly and/or domestic consumers, nationalization of declining sectors to socialize losses, and other state policies. For this group, direct state intervention in the productive sphere is not the focus of study. The CME (Capitalisme monopoliste d'Etat) approach focuses instead on how monopoly capital is furthered by state measures that transfer the formal ownerships of capital and/or redistribute profits among capitals. Although more mechanistic and economistic than the other approaches, this approach stresses the changing economic and political procedures needed to regulate capital accumulation within successive stages of capitalism (Jessop 1990).

The Amsterdam school's approach is based on a Marxist critique of political economy and a Gramscian analysis of hegemonic strategies (Bode 1979; Holman 1989; Overbeek 1988; van der Pijl 1984, 1988, 1989). This school's key concepts are fractions of capital (especially money versus productive capital) and "comprehensive concepts of control." The latter refers to the hegemonic project intended to win both bourgeois and popular support and grounded in an accumulation strategy that advances the specific interests of the dominant fraction but also secures the needs of capital in general and provides a flow of material and/or symbolic rewards to the subordinate classes. Fractions of capital are analyzed on different levels of abstraction: (1) the capital-labor relation, (2) the circulation of capital, and (3) the distribution of profit. Economic and political class strategies are analyzed on all three levels of abstraction corresponding to the comprehensive concepts of control characteristic of specific historical regimes. The comprehensive concepts of control serve to unify the ruling class and attract mass support in a hegemonic project that is needed to secure the conditions for capital accumulation and political class domination (Jessop 1990).

The West German school is best known for the works of Joachim Hirsch and his students (Hirsch and Roth 1987; Haeusler and Hirsch 1987) but contains other works also (Lutz 1984). The most distinctive feature of this school's work is the concept of "societalization," which implies the regulation of capitalist societies as a whole, not just the capitalist economy. "Societalization" regulates societies through specific modes of mass integration and the formation of a historic bloc that unifies the economic base and its political and ideological superstructures. One of the distinctive features of this approach is the concern with the role of the state and political parties in securing the conditions for effective societal regulation (Jessop 1990).

The Nordic school is primarily associated with the social scientists who have participated in the Nordic Economic Policy Project, which contrasted "economic policy models" in Nordic countries (Andersson 1984; Mjoset 1986, 1987; Mjoset et al. 1988). This approach is similar to the Parisian school but focuses on the dominant export sector as it impacts national modes of growth and national modes of policy-making that reflect the mode of growth, the political traditions, and the changing balance of economic and political forces in each country (Jessop 1990).

16. Lipietz presents two plausible resolutions to the crisis of Fordism: a productivist competitive model based on flexible automated electronics sometimes called neo-Fordism and a model of development grounded in the mass extension of nonwage-earning activity such as leisure, creativity, and intellectual enrichment. Just before his death futurologist Herman Kahn identified the U.S. recovery of 1983 with the takeoff of the new Kondratieff cycle, which is based on the emergence of new sectors formed around the automation-electronics revolution. He also predicted the formation of a new center of world economics in the Pacific rim. Lipietz argues that the "virtuous configuration" by which Japan and the Asiatic NICs are tied to the "rocket" of the U.S. recovery is exceptionally "brittle," as it is "based on immense public debt" in the United States and Japan and "it is only maintained by pumping capital out of the rest of the world," which "has limited the European recovery, maintains Latin America in depression, and plunges whole areas of the Third World into famine and death. No virtuous global cycle has yet emerged" (Lipietz 1987a, 50).

Lipietz chooses to focus on the social links between technology and a model of development. He is concerned with questions such as Who decides the immediate relations of production? How is collective labor organized? Will there be enough consumers and investors? What is produced? and How will it ensure a form of full employment? In other words, a new regime of accumulation and associated mode of regulation that provides for a consensus on the characteristic of the "wage relation" have yet to be invented. To further complicate the matter, the new regime and mode "must be compatible with a 'new international configuration' " (Lipietz 1987a, 51).

Indeed, the robotized workshop can adapt itself to changing consumer tastes and fluctuating demand by reconfiguring for short production series. Just how this impacts the post-Fordist reorganization of the labor process has been historically illustrated through two different strategies. The path chosen by most U.S. and some European (British and French) firms has been to use the automation of the management of the production process to further increase the separation between the theoretical conception of the work process and the actual implementation of work by the workers, i.e., an extension of Taylorism. The path chosen by the Japanese and some European (German and Italian) firms has been to produce a "partial requalifying of the workers, the practical knowledge of the operators being mobilized in real time in the actual automation process" (Lipietz 1987a, 51). These two paths form "the field of a tremendous social conflict, whose stake is the negotiation of a new social compromise between the involvement of workers and the apportionment of the new productivity gains" (Lipietz 1987a, 51).

That brings up the question of whom and what these productivity gains will serve. The first path mobilizes even more fixed capital per capita than did Fordism and reserves the productivity gains for profit and not for the growth of demand. The flexibility of the automated process creates the possibility that profitability will be obtained from massive investments based on short production series targeted for a well-off segment of consumers. A problem emerges if these productivity gains are not matched by mass consumption as well as elite consumption. Without mass consumption, unemployment will grow and society will be divided into three segments: a dominant class benefiting from the new products of the techno-electronic revolution, a stable but limited core of permanent wage earners, and a growing mass of uncertain workers under a weakened social security system. This fits the path chosen by the United States (Lipietz 1987a, 52).

The other path consists of a renegotiation of productivity gains with mass access to consumption goods such as home computers. This path calls for an apportionment of productivity gains in the shape of a massive reduction in work-time. This path would not be very competitive in terms of hourly wage costs relative to the model based on the continued intensification of labor. The outcome of the pursuit of both paths could possibly be the emergence of an unending struggle for hegemony leading to a permanent trade war and the "conflictual disintegration of the world economy" (Lipietz 1987a, 52). Or an outcome could obtain in which international institutional forms would be put into place that would permit the joint adoption by interested parties of some variant of the second option or at least permit a de-linking of those nations or communities who wanted to adopt the second path relative to those who choose the productivist path.

For Lipietz, the IMF's handling of the Third World debt crises provides some indication of the more likely international scenario. Though the 1982 NIC debt crisis was temporarily resolved via an average three-year moratorium on debt, the IMF's imposition of export enhancement programs and associated social austerity programs pushes the newly emerged "middle classes back into poverty, the workers towards wretchedness, and the poor towards death" (Lipietz 1987a, 52). This path resonates with Jack London's prophecies in *The Iron Heel,* where perpetual indebtedness would keep the industrialized Third World in the role of sweatshop for the First World.

The other alternative involves the cancellation of the Third World debt, similar to that of the German and Russian debts between the World Wars, or the consolidation and permanent freezing of the loaned amounts. As a result, the free issue and distribution of international currency could then provide the peripheral Fordist countries with a more autonomous road and a more self-centered regime of accumulation. The two paths represent a "more repressive" and a "more social-democratic" option. We are in a position similar to just before World War II, when Nazism or the New Deal were the choices (Lipietz 1987a, 53).

17. Habermas (1981) argues that the shift in political forms is because the new social movements' participants are not motivated by issues of distribution but by issues such as quality of life and human rights. They are critical of highly bureaucratized welfare states operated by "Social-Democratic administrators of statist rationality" (Lash and Urry 1987, 220). Grannow and Offe (1982, 80) conclude that "traditional social-democracy is thus eroded, losing its 'oppositional' elements, without being able to integrate 'alternative ones.' " Toffler and Toffler (1985, 16) argue that the emergence of postindustrialism has broken up the traditional voting blocks and created a much more diverse, "fragmented electorate made up of thousands of tiny, transitory groupings rather that the large blocs manipulated by parties like the Democrats." Since the 1960s some of these new social movements have had considerable influence over U.S. politics. Vogel (1983, 20) states: "A loose coalition of middle class based consumer and environmental, feminist and civil rights organizations, assisted on occasions by organized labor, aided by a sympathetic media and supported by much of the intelligentsia, were able to influence both the terms of public debate and the outcomes of government policy in a direction antithetical to the interest of business."

CHAPTER 4. THE TUNA FISHING INDUSTRY IN THE FORDIST PERIOD

1. The terms porpoise and dolphin refer to the same animals. In general porpoise is the term most often used by fishermen, while dolphin is the more scientific term.

CHAPTER 5. THE RESTRUCTURING OF THE GLOBAL TUNA INDUSTRY

1. The majority of landings in the American Samoa/California category are in American Samoa. The category is used to maintain confidentiality.

CHAPTER 6. LAWSUITS, COMPROMISES, AND EMBARGOES

1. EII was founded in 1982 by David Brower of Sierra Club and Friends of the Earth.

2. In March 1991 Mark A. Koob was promoted to CEO of Bumble Bee. Under Koob, Bumble Bee passed Chicken of the Sea to become second in the U.S. tuna market, with $400 million in sales. Market share in 1990 was Star Kist, 35.5 percent; Bumble Bee, 20.2 percent; and Chicken of the Sea, 18.5 percent. Koob said that his company had gained sales momentum the previous year after introducing a new loin technology that allowed it to ship frozen tuna to a canning plant near San Diego. The process saves money on shipping costs, allowing Bumble Bee to lower some prices and avoid high tariffs on canned tuna versus bulk tuna. Thirty-seven-year-old Koob worked for Bumble Bee from 1982–1987 and has also worked for Proctor and Gamble, Kimberly-Clark Corporation, and Ralston Purina (*New York Times* 1991).

CHAPTER 7. GLOBAL POST-FORDISM: RESTRUCTURING AND THE CONDITIONS OF SOLIDARITY

1. The concept of the production process is employed here in its broader meaning. It includes all the operations and strategies that lead to the creation of the final product.

2. In effect, this is an argument that follows the regulationists' thesis. They view as not possible any form of systemic organization outside national social formations. Once the range of economic operations is transnationalized, systemic disorganization is generated.

3. To be sure, it would be erroneous to equate the establishment of "free spaces" with formal democracy. As illustrated by the voluminous literature in political sociology (e.g., Habermas 1975; Marcuse 1964; Offe 1985), the availability of democratic spaces within the sphere of the state is the outcome of struggles and negotiations among various social groups that establish the separation between the state's formal and substantive dimensions.

4. The emergence of a transnational state must address the spacial distancing between production (i.e., global investment, global allocation of production tasks, reallocation of employment opportunities across national boundaries, and so on) and social demands (i.e., employment opportunities, availability of consumer items, social services, and so on). More specifically, both TNCs and subordinate classes must confront the issue of the discrepancy between the needs of people and regional economic development, which remain local, and the economic and political processes that can address them, which increasingly tend to be located in the transnational arena. In terms of TNCs, this situation mandates the existence and availability of instruments of social control and social legitimation that might minimize the emergence of social instability. Moreover, it mandates the creation of instruments that can maintain and/or expand the production and

consumption capacities essential for continuous capital accumulation. For subordinate classes, this process involves an increasing inability to resort to traditional channels of action for the satisfaction of economic and social needs. In terms of employment possibilities, the attraction of new jobs is often a function of discounted wages and community investments in favor of incoming enterprises. The maintenance of existing jobs is threatened by economic claims and by the very process of attracting new investments, which demands the taxing of existing activities. In terms of social services, the distancing of the economic sphere from the sociopolitical sphere has emerged as one of the most relevant factors exacerbating the fiscal crisis of the local state.

5. Accordingly, this case study speaks directly against the neoliberal assumption maintaining that unregulated capitalism can successfully address economic growth, employment, and protection of the environment. TNCs' constant attempts to avoid proenvironment legislation, the waves of negative consequences for labor in the United States and Latin America, and the use of global sourcing by TNCs to avoid the task of developing environmentally sound fishing technologies point out the limits of the neoliberal proposal.

6. This should not be interpreted as a stand against diversity. On the contrary, it speaks to the fragmenting of common experience that in the past constituted the backbone of collective movements such as unions and political organizations of the working class.

7. This does not imply that the labor movement and the environmental movement are progressive by definition. In fact, these are complex movements that include a number of fractions inspired by diverging ideologies. However, their histories are characterized in general by antiestablishment postures. Moreover, arguably the most significant of their ideological expressions are grounded in critiques of capitalism.

8. The literature discussed in Chapter 2 provides more than sufficient evidence of the growing class polarization of today's society.

BIBLIOGRAPHY

Aglietta, Michel. 1988. "Regimes monetaires, monnaie internationale, monnaie commune." Paper presented at the International Conference on Regulation Theory, Barcelona, Spain.

——. 1986. *Le Declin des devises cles.* Paris: La Decouverte.

——. 1982. "World capitalism in the eighties." *New Left Review* 136:5–42.

——. 1979. *A Theory of Capitalist Regulation.* London: Verso.

——. 1978. "Internationalisation des relations financieres et de la production." In *Economie internationale,* ed. J. Weiller and J. Coussy, vol. 2, *Internationalisation et integration ou co-operation,* 32–65. Paris: INSEE.

——. 1975. "Problematique de la valeur." In *Economie internationale,* ed. J. Weiller and J. Coussy, vol. 1, *Automatisation et structures,* 34–75. Paris: INSEE.

——. 1974. *Accumulation et regulation du capitalisme en longue periode: Exemple des Etats-Unis (1870–1970).* Paris: INSEE.

Aglietta, Michel, and Anton Brender. 1984. *Les metamorphoses de la societe salariale: La France en project.* Paris: Calmann-Levy.

Akard, Patrick J. 1992. "Corporate mobilization and political power: The transformation of U.S. economic policy in the 1970s." *American Sociological Review* 57(5): 597–615.

Allen, Francis R. 1974. *Socio-Cultural Dynamics.* New York: Macmillan.

Anderson, Ian. 1988. "Government observers 'forced to lie' about dolphin deaths." *New Scientist,* June 16:34.

Andersson, J. O. 1984. "The state in the transformation of industrially mature societies." In *The Role of the State in Social Transformation Under the Impact of World Crisis,* ed. K. Kulcsar, 142–59. Budapest: University Press.

Antonio, Robert J., and Alessandro Bonanno. 1995. "Post-Fordism: The poverty of market centered democracy." In *Current Perspectives in Social Theory* (forthcoming).

Asada, Yohoji. 1972. "Management and investment in Japan's tuna longline fishery." In *Economic Aspects of Fish Production,* ed. OECD, 309–18. Paris: Organization for Economic Cooperation and Development.

Ashender, G. P., and G. W. Kitson. 1987. "*Japanese Tuna Fishing/Processing Companies.*" Honolulu: Pacific Islands Development Program, East-West Center.
Audubon. 1988. "Porpoise mortality numbers skewed." 90(5): 16.
Babcock, Blair. 1984. *Unfairly Structured Cities.* Oxford: Basil Blackwell.
Badie, Bertrand, and Pierre Birnbaum. 1983. *The Sociology of the State.* Chicago: University of Chicago Press.
Bailey, Conner. 1988. "The political economy of fisheries development in the Third World." *Agriculture and Human Values* 5(1–2): 35–48.
Barlett, Donald L., and James B. Steele. 1992. *America: What Went Wrong.* Kansas City, Mo.: Andrews and McMeel.
Baudrillard, Jean. 1983. *In the Shadow of the Silent Majority.* New York: Semiotext(e).
Baulant, C. 1988. "Role de l'etat chez les theoriciens de la regulation." Paper presented at the International Conference on Regulation Theory, Barcelona, Spain.
Bell, Daniel. 1976. *The Coming of Post-Industrial Society: A Venture in Social Forcasting.* New York: Basic Books.
Ben-Yami, M. 1980. *Tuna Fishing with Pole and Line.* Food and Agricultural Organization of the United Nations. Surrey, England: Fishing News Books.
Black, Dorothy. 1992. "International trade v. environmental protection: The case of the U.S. embargo of Mexican tuna." *Columbia Law Review* 24(1): 123–56.
Block, Fred. 1980. "Beyond relative autonomy: State managers as historical subjects." In *Socialist Register,* ed. R. Miliband and J. Seville, 227–40. London: Merlin Press.
———. 1977. "The ruling class does not rule." *Socialist Revolution* 7(3): 6–28.
Bluestone, Barry, and Bennett Harrison. 1986. "The great American job machine: The proliferation of low wage employment in the U.S. economy." Study prepared for the Joint Economic Committee, U.S. Congress, December.
———. 1982. *The Deindustrialization of America: Plant Closing, Community Abandonment, and the Dismantling of Basic Industry.* New York: Basic Books.
Boccara, Paul. 1985. *Intervenir dans les gestions avec de nouveaux criteres.* Paris: Messidor/Editions sociales.
———. 1983. "Cycles longs: Mutations technologiques et originalite de la crise de structure actuelle." *Issues* 16:5–60.
Bode, R. 1979. "De Nederlandse bourgeoisie tussen de twee wereldoor-logen." *Cahiers voor de politieke en sociale Wetenschappen* 2(4): 9–50.
Bonanno, Alessandro. 1994. "The locus of polity action in a global setting." In *From Columbus to ConAgra: The Globalization of Agriculture and Food,* ed. Bonanno, Busch, et al., 251–64. Lawrence: University Press of Kansas.
———. 1993. "The agro-food sector and the transnational state: The case of the EC." *Political Geography* 12(4): 341–60.
———. 1992. "Globalization of the agricultural and food sector: The crisis of contradictory convergence." In *The Agricultural and Food Sector in the New Global Era,* ed. A. Bonanno, 25–50. London: Concept Publishing.
———. 1991. "The globalization of the agricultural and food system and theories of the state." *International Journal of Sociology of Agriculture and Food* 1:15–30.

———. 1990. "Introduction." In *Agrarian Policies and Agricultural Systems,* ed. A. Bonanno, 1–8. Boulder, Colo.: Westview Press.

———. 1988. "Theories of the state: The case of land reform in Italy." *Sociological Quarterly* 29(1): 131–47.

———. 1987a. "Agricultural policies and the capitalist state." *Agriculture and Human Values* 4(2–3): 40–46.

———. 1987b. *Small farms: Persistence with legitimation.* Boulder, Colo.: Westview Press.

Bonanno, Alessandro, Lawrence Busch, William H. Friedland, Lourdes Gouveia, and Enzo Mingione, eds. 1994. *From Columbus to ConAgra: The Globalization of Agriculture and Food.* Lawrence: University Press of Kansas.

Bonanno, Alessandro, and Douglas H. Constance. 1993. "The state and the regulatory process: The case of the global tuna-fish industry." Paper presented at workshop, "Concepts of the state in a changing global agri-food system," Wageningen, Holland, July 31–August 2, 1993.

Bonanno, Alessandro, Donato Fernandez, and Jere L. Gilles. 1990. "Agrarian policies in the U.S. and EC: A comparative analysis." In *Agrarian Policies and Agricultural Systems,* ed. A. Bonanno, 227–51. Boulder, Colo.: Westview Press.

Bonanno, Alessandro, J. Sanford Rikoon, Douglas Constance, Mary Hendrickson, and Paula Owsley. 1993. "The global agri-food sector and limits to the regulation and protection of the environment: The case of the U.S. poultry industry." University of Missouri–Columbia. Manuscript.

Borrego, John. 1981. "Metanational capitalist accumulation and the emerging paradigm of revolutionist accumulation." *Review* 4(4): 713–77.

Bowles, Samuel, and Herbert Gintis. 1982. "The crisis of liberal democratic capitalism: The case of the United States." *Politics and Society* 11:52–92.

Bowles, Samuel, David M. Gordon, and Tom Weisskopf. 1988. "Social institutions, interests, and the empirical analysis of accumulation: A reply to Bruce Norton." *Rethinking Marxism* 1: 44–57.

Boyer, Robert. 1986a. *Capitalismes fin de siecle.* Paris: PUF.

———. 1986b. *La Flexibilite du travail en Europe. Une Etude comparative des transformations dur rapport salarial dans sept pays.* Paris: La Decouverte. Translated as *The Search for Labour Market Flexibility: The European Economies in Transition.* Oxford: Clarendon, 1988.

———. 1986c. *La Theorie de la regulation: Une Analyze critique.* Paris: la Decouverte.

———. 1985. "Analyses de la crise Americaine: A propos d'un ouvrage recent." Paris: CEPREMAP 8501.

———. 1979. "Wage formation in historical perspective: The French experience." *Cambridge Journal of Economics* 3:99–118.

Boyer, Robert, and Samuel Bowles. 1988. "Labor discipline and aggregate demand: A macroeconomic model." *American Economic Review* 78(2): 395–400.

Boyer, Robert, and Jacques Mistral. 1984. "Entre l'etat et le march: Conjuguer l'economique et le social." *Esprit* 1:109–28.

Bradbury, Katherine. 1986. "The shrinking middle class." *New England Economic Review* (September/October): 47.

Bradsher, Keith. 1992. "U.S. told to bar tuna over dolphin killings." *New York Times,* January 15, 141:D16.

Brooke, James. 1992a. "America—environmental dictator?" *New York Times,* May 3, 141:F7.

———. 1992b. "10 nations reach accord on saving dolphins." *New York Times,* May 12, 141:C4.

Brower, Kenneth. 1989. "The destruction of dolphins: In spite of laws intended to protect them, federal indifference and cruel fishing methods once again endanger dolphins." *Atlantic* 263(1): 35.

Bullock, Bruce. 1991. "Research challenges for social scientists generated by globalization." Paper presented at the International Conference on the Globalization of the Agricultural and Food Order, University of Missouri–Columbia, June.

Burton, Johnathan. 1989. "Lease of shelf life." *Far East Economic Review* 145(36): 108.

Busch, Lawrence. 1994. "The state of agricultural science and the agricultural science of the state." In *From Columbus to ConAgra: The Globalization of Agriculture and Food,* ed. Bonanno, Busch, et al., 69–84. Lawrence: University Press of Kansas.

Busch, Lawrence, William B. Lacy, Jeffrey Burkhardt, and Laura Lacy. 1991. *Plants, Power, and Profit.* Oxford: Basil Blackwell.

Business Week. 1990. "The stateless corporation." May 14:98–105.

Buttel, Frederick H. 1994. "Agricultural change, rural society, and the state in late twentieth century: Some theoretical observations." In *Agricultural Restructuring and Rural Change in Europe,* ed. David Symes and Anton J. Jansen, 13–31. Wageningen, Holland: Agricultural University Wageningen.

———. 1992. "Environmentalization: Origins, processes, and implications for rural social change." *Rural Sociology* 57(1): 1–27.

Calasanti, Toni M., and Alessandro Bonanno. 1992. "Working 'overtime': Economic restructuring and retirement of a class." *Sociological Quarterly* 33(1): 135–52.

Campbell, John L., and Leon N. Lindberg. 1990. "Property rights and the organization of economic activity by the state." *American Sociological Review* 55(5): 634–47.

Carnoy, Martin. 1984. *The State and Politial Theory.* Princeton: Princeton University Press.

Chandler, Alfred D. 1977. *The Visible Hand: The Managerial Revolution in American Business.* Cambridge, Mass.: Belknap Press.

Clarke, Simon. 1990. "The crisis of Fordism or the crisis of social-democracy?" *Telos* 83:71–98.

———. 1988. "Overaccumulation, class struggle, and the regulation approach." *Capital and Class* 36:59–92.

Coe, James, and George Sousa. 1972. "Removing porpoise from a tuna purse seine." *Marine Fisheries Review.* Reprint no. 949: 15–19.

Colson, David A. 1992. "U.S. policy on tuna-dolphin issues." *United States Department of State Dispatch* 3(340): 667–71.

Comitini, S. 1987. "Japanese trading companies: Their possible role in Pacific tuna fisheries development." Honolulu: Pacific Islands Development Program, East-West Center.

Constance, Douglas H., Alessandro Bonanno, and William D. Heffernan. 1993. "Global contested terrain: The case of the tuna-dolphin controversy." Paper presented at the annual meeting of the Rural Sociology Society, Orlando, Florida, August.

Constance, Douglas H., Jere L. Gilles, and William D. Heffernan. 1990. "Agrarian policies and agricultural systems in the United States." In *Agrarian Policies and Agricultural Systems,* ed. Alessandro Bonanno, 9–75. Boulder, Colo.: Westview Press.

Constance, Douglas H., and William D. Heffernan. 1991. "The global poultry agro/food complex." *International Journal of Sociology of Agriculture and Food* 1:126–42.

Cox, Robert W. 1981. "Social forces, states, and world orders: Beyond international relations theory." *Millennium* 10:126–55.

Crough, G. J. 1987. "Australian tuna industry." Honolulu: Pacific Islands Development Program, East-West Center.

Dahrendorf, Ralph. 1987. "The erosion of citizenship and its consequences for us all." *New Statesman,* June 12.

———. 1959. *Class and Class Conflict in Industrial Societies.* Berkeley: Stanford University Press.

Danziger, Sheldon, Peter Gottschalk, and Eugene Smolensky. 1989. "How the rich have fared." *American Economic Review* 90:311.

Davis, Andrew. 1988. "Caught in the tuna nets: the slaughter of dolphins." *The Nation,* November 14, 247(14): 486.

Davis, Bob. 1992. "U.S., Mexico, Venezuela set accord on tuna." *Wall Street Journal,* March 20:B10(E).

Davis, Kingsley, and Wilbert E. Moore. 1945. "Some principles of stratification." *American Sociological Review* 10(3): 242–49.

Delorme, Robert. 1987. "State intervention in the economy revisited: An outline." Paris: Unpublished paper.

Delorme, Robert, and Christine Andre. 1982. *L'Etat et l'economie.* Paris: Seuil.

Devine, Joel. 1985. "State and state expenditures: Determinants of social investment and social consumption spending in the postwar United States." *American Sociological Review* 50(2): 150–65.

Domhoff, William G. 1979. *The Powers That Be.* New York: Vintage Books.

———. 1967. *Who Rules America?* Englewood Cliffs, N.J.: Prentice-Hall.

Doulman, David J. 1987. "Distant-water fleet operations and regional fisheries cooperation." In *The Development of the Tuna Industry in the Pacific Islands Region: An Analysis of Options,* ed. David J. Doulman, 33–52. Honolulu: East-West Center.

Drugman, B. 1984. "A nouveau sur la question de la regulation. Economie poitique, marxisme et . . . crise: quelle alternative reelle?" *Economies et Societes* 18(11): 29–64.

The Economist. 1991. "Divine porpoise." October 5, 321(7727): A31.

Ehrenreich, Barbara. 1989. *Fear of Falling: The Inner Life of the Middle Class.* New York: Pantheon Books.

Ellison, Katherine. 1991. "U.S. quest for dolphin-safe tuna hurts Mexican fisherman." *Journal of Commerce and Commercial* 390(27592): 5A.

Facts on File. 1992. "Yellowfin tuna ban enforced." February 6, 52(2672): 78.

Fernandez, Linda, and Linda Lucas Hudgins. 1987. "A summary of international tuna markets: Characteristics and accessibility for Pacific island countries." In *The Development of the Tuna Industry in the Pacific Islands Region: An Analysis of Options,* ed. David J. Doulman, 135–52. Honolulu: East-West Center.

Fine, Ben, and Lawrence Harris. 1979. *Rereading Capital.* London: Macmillan.

Flint, Jerry. 1992. "Honda the most efficient 'American' carmaker." *Forbes* 149(1): 121.

Florida, Richard, and Marshall M. A. Feldman. 1988. "Housing in U.S. Fordism: The class accord and postwar spatial organization." *International Journal of Urban and Regional Research* 12(2): 187–210.

Floyd, Jesse M. 1987. "U.S. tuna import regulations." In *The Development of the Tuna Industry in the Pacific Islands Region: An Analysis of Options,* ed. David J. Doulman, 81–90. Honolulu: East-West Center.

Friedland, William H. 1994. "The new globalization: The case of fresh produce." In *From Columbus to ConAgra: The Globalization of Agriculture and Food,* ed. Bonanno, Busch, et al., 210–31. Lawrence: University Press of Kansas.

——. 1991a. "Introduction." In *Towards a New Political Economy of Agriculture,* ed. William H. Friedland, Lawrence Busch, Frederick H. Buttel, and Alan P. Rudy, 1–34. Boulder, Colo.: Westview Press.

——. 1991b. "The transnationalization of agricultural production: Palimpsest of the transnational state." *International Journal of Sociology of Agriculture and Food* 1:48–58.

——. 1984. "Commodity systems analysis: An approach to the sociology of agriculture." In *Research in Rural Sociology and Development: A Research Annual,* ed. Harry K. Schwarzweller, 221–35. Greenwich, Conn.: JAI Press.

——. 1983. "State formation and reformation in California grapes." Paper presented at conference, The Political Economy of Agriculture, Ann Arbor, Michigan, August.

Friedmann, Harriet, and Philip McMichael. 1989. "Agriculture and the state system." *Sociologia Ruralis* 29(2): 93–117.

Galbraith, John Kenneth. 1979. *The New Industrial State.* 3d ed. rev. New York: New American Library.

Gamble, Andrew. 1988. *The Free Economy and the Strong State: The Politics of Thatcherism.* Basingstoke, England: MacMillan.

Giardina, Denise, and Eric Bates. 1991. "Fowling the nest." *Southern Exposure* 19(1): 8–12.

Giddens, Anthony. 1990. *The Consequences of Modernity.* Stanford: Stanford University Press.

——. 1987. *Social Theory and Modern Sociology.* Cambridge, England: Polity Press.

Gilbert, Jess, and Carolyn Howe. 1991. "Beyond 'state vs. society': Theories of the state and New Deal agricultural policies." *American Sociological Review* 56(2): 204–20.

Gill, Stephan. 1990. *American Hegemony and the Trilateral Commission.* Cambridge: Cambridge University Press.

Godges, John. 1988. "Dolphins hit rough seas again." *Sierra* 73(3): 24.

Gordon, David M. 1988. "The global economy: New edifice or crumbling foundations." *New Left Review* 168:24–65.

——. 1980. "Stages of accumulation and long economic cycles." In *Processes of the World System,* ed. T. Hopkins and I. Wallerstein, 72–94. Beverly Hills, Calif.: Sage.

——. 1978. "Up and down the long roller coaster?" In *U.S. Capitalism in Crisis,* ed. Union for Radical Political Economics, 22–35. New York: URPE.

Gordon, David M., Richard Edwards, and Michael Reich. 1982. *Segmented Work, Divided Workers.* Cambridge: Cambridge University Press.

Gorz, Andre. 1982. *Farewell to the Working Class: An Essay on Post-Industrial Socialism.* London: Pluto Press.

Gottdiener, Mark. 1993. "A Marx for our time: Henri Lefebvre and *The Production of Space.*" *Sociological Theory* 11(1): 129–34.

——. 1985. *The Social Production of Urban Space.* Austin: University of Texas Press.

Gottdiener, Mark, and Nikos Komninos, eds. 1989. *Capitalist Development and Crisis Theory: Accumulation, Regulation, and Spatial Restructuring.* London: Macmillan.

Gouveia, Lourdes. 1994. "Global strategies and local linkages: The case of the U.S. meatpacking industry." In *From Columbus to ConAgra: The Globalization of Agriculture and Food,* ed. Bonanno, Busch, et al., 125–48. Lawrence: University Press of Kansas.

——. 1992. "Immigrant labor in the internationalization of meat processing." In *The Agricultural and Food Sector in the New Global Era,* ed. Al. Bonanno, 82–101. London: Concept Publishing.

Gramsci, Antonio. 1975. *Quaderni dal Carcere.* Rome: Editori Riuniti.

——. 1971. *Selection from Prison Notebooks.* New York: International Publishers.

Grannow, V., and Claus Offe. 1982. "Political culture and the politics of the Social Democratic government." *Telos* 53:67–80.

Green, Gary. 1987. "The political economy of flue-cured tobacco production." *Rural Sociology* 52(2): 221–41.

Green, R. E., W. F. Perrin, and B. P. Petrich. 1971. "The American tuna purse-seine fishery, "In *Modern Fishing Gear in the World,* ed. Hilmar Kristjonnson, 182–94. London: Fishing News (Books).

GRREC. 1983. *Crise et regulation: Recueil de textes, 1979–1983.* Grenoble: DRUC.

Habermas, Jürgen. 1987. *The Theory of Communicative Action.* 2 vols. Boston: Beacon Press.

——. 1985. *Legitimation Crisis.* Boston: Beacon Press.

——. 1981. "New social movements." *Telos* 49(181): 33–37.

Haeusler, J., and J. Hirsch. 1987. "Regulation und parteien im uebergang zum 'post-Fordismus.' " *Das Argument* 165:651–71.

Handley, Paul. 1991a. "Off the hook." *Far East Economic Review,* May 23, 151(21): 48–49.

——. 1991b. "Row of canneries." *Far East Economic Review,* May 23, 151(21): 50.

——. 1989. "Unicord's big catch." *Far East Economic Review,* September 7, 145(36): 108–9.

Harrison, Bennett, and Barry Bluestone. 1988. *The Great U-Turn: Corporate Restructuring and the Polarizing of America.* New York: Basic Books.

——. 1983. *Beyond the Wasteland.* New York: Doubleday.

Harvey, David. 1990. *The Condition of Postmodernity.* Oxford: Basil Blackwell.

Hechter, Michael, and William Brustein. 1980. "Regional modes of production patterns and state formation in Western Europe." *American Sociological Review* 85(5): 1061–94.

Heffernan, William D. 1990. "The transnationalization of the poultry industry." Paper presented at the Twelfth World Congress of Sociology, Madrid, Spain, July.

Heffernan, William D., and Douglas H. Constance. 1994. "Transnational Corporations and the Global Food System." In *From Columbus to ConAgra: The Globalization of Agriculture and Food,* ed. Bonanno, Busch, et al., 29–51. Lawrence: University Press of Kansas.

Heilbroner, Robert. 1993. *The Making of Economic Society.* Englewood Cliffs, N.J.: Prentice-Hall.

Heinz, H. J. 1991. *Annual Report.* Omaha, Nebr.: Heinz Corporation.

——. 1985. *Annual Report.* Omaha, Nebr.: Heinz Corporation.

Held, David. 1983. "Central perspectives on the modern state." In *States and Societies,* ed. David Held, 67–89. New York: New York University Press.

Herrick, Samuel F., Jr., and Steven J. Koplin. 1986. "U.S. tuna trade summary, 1984." *Marine Fisheries Review* 48(3): 28–37.

Hirsch, J., and R. Roth. 1987. *Das neue Gesicht des Kapitalismus.* Hamburg, Germany: VSA Verlag.

Hirsch, Paul, and Jonathan Zeitlan. 1991. "Flexible specialization versus post-Fordism: Theory, evidence, and policy implications." *Economy and Society* 20(1): 1–56.

——. 1988. *Reversing Industrial Decline.* Oxford: Berg.

Hodgson, Godfrey. 1978. *America in Our Time.* New York: Vintage Books.

Hofman, R. J., ed. 1981. *Identification and Assessment of Possible Alternative Methods for Catching Yellowfin Tuna.* Sponsored by the Marine Mammal Commission in cooperation with the Inter-American Tropical Tuna Commission and the National Marine Fisheries Service, Report no. PB83-138933. Springfield, Va.: National Technical Information Service.

Holland, Kerry L. 1991. "Exploitation on porpoise: The use of purse-seine nets by commercial tuna fisherman in the Eastern Tropical Pacific Ocean." *Syracuse Journal of International Law and Commerce* 17:241.

Holman, Otto. 1989. "Felipismo, or the failure of neo-liberalism in Spain." Paper prepared for the Political Studies Association Conference, Coventry, England, April.

——. 1987–1988. "Semiperipheral Fordism in southern Europe: The national and international context of socialist-led governments in Spain, Portugal, and Greece in historical perspective." *International Journal of Political Economy* 17(4): 11–55.

Hooks, Gregory. 1990. "From an autonomous to a captured state agency." *American Sociological Review* 55(1): 29–43.

Horkheimer, Max. 1974. *The Eclipse of Reason.* New York: Seabury Press.

Hoselitz, Bert F. 1960. *Theories of Economic Growth.* Glencoe, Ill.: Free Press.

Hudgins, Linda Lucas. 1987. "The development of the tuna industry in Mexico: 1976–1986." In *The Development of the Tuna Industry in the Pacific Islands Region: An Analysis of Options,* ed. David J. Doulman, 153–68. Honolulu: East-West Center.

Hudgins, Linda Lucas, and Linda Fernandez. 1987. "A summary of international business operations in the global tuna market." In *The Development of the Tuna Industry in the Pacific Islands Region: An Analysis of Options,* ed. David J. Doulman, 289–301. Honolulu: East-West Center.

Hymer, Stephen. 1976. *The International Operation of National Firms: A Study of Direct Foreign Investment.* Cambridge, Mass.: MIT Press.

——. 1971. "The multinational corporation and the international division of labor." In *The Multinational Corporation: A Radical Approach,* ed. Robert B. Cohen, 47–65. Cambridge: Cambridge University Press.

Hymer, Stephen, and Stephen A. Resnick. 1971. "International trade and uneven development." In *The Multinational Corporation: A Radical Approach,* ed. Robert B. Cohen, 66–83. Cambridge: Cambridge University Press.

IATTC (Inter-American Tropical Tuna Commission). 1989. "The fishery for tunas in the Eastern Pacific Ocean." Tuna-Dolphin Workshop, San Jose, Costa Rica, March 14–16, 1989. Working Document no. 1. La Jolla, Calif.: Scripps Institute of Oceanography.

Inkeles, Alex, and David Smith. 1974. *Becoming Modern: Individual Change in Six Developing Countries.* Cambridge: Harvard University Press.

Iversen, Robert T. B. 1987a. "U.S. tuna processors." In *The Development of the Tuna Industry in the Pacific Islands Region: An Analysis of Options,* ed. David J. Doulman, 271–88. Honolulu: East-West Center.

——. 1987b. "U.S. tuna processors." *Pacific Islands Development Program.* Honolulu: East-West Center.

Jameson, Fredric. 1984. "Postmodernism, or the cultural logic of late capitalism." *New Left Review* 146:53–92.

Jenkins, Craig, and Barbara Brents. 1989. "Social protest, hegemonic competition, and social reform." *American Sociological Review* 54(6): 891–909.

Jessop, Bob. 1990. "Regulation theories in retrospect and prospect." *Economy and Society* 19(2): 153–216.

——. 1983. *The Capitalist State.* Oxford: Basil Blackwell.

——. 1982. "Accumulation strategies, state forms, and hegemonic projects." *Kapitalistate* 10/11: 89–112.

Joseph, James. 1986. *Recent Developments in the Fishery for Tropical Tunas in the Eastern Pacific Ocean.* La Jolla, Calif.: Inter-American Tropical Tuna Commission.

——. 1973. "The scientific management of world stocks of tunas, billfishes and related species." FAO-Fi-FMD/73/S-48.

——. 1972. "An overview of tuna fisheries of the world." In *Economic Aspects of Fish Production.* Paris: Organization for Economic Cooperation and Development.

Joseph, James, and Joseph W. Greenough. 1979. *International Management of Tuna, Porpoise, and Billfish.* Seattle: University of Washington Press.

Journal of Commerce and Commercial. 1991. "Tight-lipped about tuna." 390(27577): 4A.

Kenney, Martin, Linda M. Lobao, James Curry, and Richard Coe. 1989. "Midwest agriculture and U.S. Fordism." *Sociologia Ruralis* 29(2): 131–48.

Kindleberger, Charles P. 1986. "International public goods without international government." *American Economic Review* 76(1): 1–13.

King, Dennis M. 1987. "The U.S. tuna market: A Pacific Islands perspective." In *The Development of the Tuna Industry in the Pacific Islands Region: An Analysis of Options,* ed. David J. Doulman, 63–80. Honolulu: East-West Center.

Kneen, Brewster. 1990. *Trading Up: How Cargill, the World's Largest Grain Company, Is Changing Canadian Agriculture.* Toronto: NC Press.

Koc, Mustafa. 1994. "Globalization as a discourse." In *From Columbus to ConAgra: The Globalization of Agriculture and Food,* ed. Bonanno, Busch, et al., 265–80. Lawrence: University Press of Kansas.

——. 1991. "Globalization, compartmentalization and the new world order." Paper presented at the International Conference on the Globalization of the Agricultural and Food Order, University of Missouri–Columbia, June.

Kraul, Chris. 1990. "U.S. fisherman fear decision may be final blow." *Los Angeles Times,* April 14, 109:D1.

——. 1989. "Pillsbury to sell Bumble Bee unit to Thai firm." *Los Angeles Times,* August 17, 108:2.

Krebs, A. V. 1992. *The Corporate Papers: The Book of Agribusiness.* Washington, D.C.: Essential Books.

Kronman, Mick. 1991. "Fishing morally correct tuna." *Journal of Commerce and Commercial* 390(27613): 8A.

Laclau, Ernesto, and Chantal Mouffe. 1984. *Hegemony and Socialist Strategy: Toward a Radical Democratic Politics.* London: Verso.

Lambert, John. 1991. "Europe: The nation-state dies hard." *Capital & Class* 43(2): 9–23.

Lash, Scott, and John Urry. 1987. *The End of Organized Capitalism.* Madison: University of Wisconsin Press.

Leinberger, Paul, and Bruce Tucker. 1991. *The New Individualists.* New York: HarperCollins.

Levin, Myron. 1989. "Dolphin demise: Foreign tuna fishing fleets blamed for most of the sharp increase in killings." *Los Angeles Times,* March 5, 108:3.

Levine, Rhonda. 1988. "Theoretical developments on the question of state autonomy: The case of New Deal industrial policies." Paper presented at the annual meeting of the American Sociological Association, Atlanta.

——. 1987. "Bringing classes back in: State theory and theories of the state." In *Recapturing Marxism,* ed. Rhonda Levine and Jerry Lembcke, 10–28. New York: Praeger Publishers.

Levine, Rhonda, and Jerry Lembcke. 1987. "Introduction: Marxism, neo-Marxism, and U.S. sociology." In *Recapturing Marxism,* ed. Rhonda Levine and Jerry Lembcke, 1–9. New York: Praeger Publishers.

Lipietz, Alain. 1992. *Towards a New Economic Order: Post-Fordism, Ecology, and Democracy.* New York: Oxford University Press.

——. 1991. "A regulationist approach to the future of urban ecology." *Capitalism, Nature, Socialism* 3(3): 101–10.

——. 1988. "Building an alternative movement in France." *Rethinking Marxism* 1(3): 80–99.

——. 1987a. "The globalization of the general crisis of Fordism." In *Frontyard Backyard: The Americas in the Global Crisis,* ed. J. Holmes and Colin Leys, 23–56. Toronto: Between the Lines.

——. 1987b. *Mirages and Miracles.* London: New Left Books.

——. 1986. "Caracteres seculaires et conjoncturels de l'intervention economique de l'etat." Paris: CEPREMAP 8621.

——. 1985. "Le national et le regional: Quelle autonomie face a la crise capitalists mondiale?" Paris: CEPREMAP 8521.

——. 1984. "Imperialism or the beast of the apocalypse." *Capital and Class* 22:81–110.

——. 1982. "Towards global Fordism." *New Left Review* 132:33–47.

Lipset, Seymour Martin. 1963. *Political Man: The Social Basis of Politics.* Garden City, N.Y.: Anchor Books.

Llambí, Luis. 1994. "Open economies and closing markets: Latin American agriculture's difficult search for a place in the emerging global order." In *From Columbus to ConAgra: The Globalization of Agriculture and Food,* ed. Bonanno, Busch, et al., 184–209. Lawrence: University Press of Kansas.

Llambí, Luis, and Lourdes Gouveia. 1994. "The restructuring of the Venezuelan state and state theory." *International Journal of Sociology of Agriculture and Food* 4:45–72.

Lobao, Linda M. 1990. *Locality and Inequality: Farm and Industry Structure and Socioeconomic Conditions.* Albany: State University of New York Press.

Lukács, George. 1971. *History and Class Consciousness.* Cambridge, Mass.: MIT Press.

Lutz, Burkart. 1984. *Der kuerze Traum immerwaehrender Prosperitaet.* Frankfurt, Germany: Campus.

Luxner, Larry. 1990. "Puerto Rico lures Asians to tuna business." *Journal of Commerce and Commercial* 384(27202): 4A.

McDorman, Ted L. 1991. "The GATT consistency of U.S. fish import embargoes to stop driftnet fishing and save whales, dolphins, and turtles." *George Washington Journal of International Law and Economics* 24:477–524.

MacEwan, Arthur, and William T. Tabb, eds. 1989. *Instability and Change in the World Economy.* New York: Monthly Review Press.

McIntyre, Richard. 1991. "The political economy and class analytics of international capital flow: U.S. industrial capital in the 1970s and 1980s." *Capital & Class* 43(2): 179–201.

McMichael, Philip. 1991. "Food, the state, and the world economy." *International Journal of Sociology of Agriculture and Food* 1:71–85.

McMichael, Philip, ed. 1994. *The Global Restructuring of Agro-Food Systems.* Ithaca and London: Cornell University Press.

McMichael, Philip, and David Myhre. 1991. "Global regulation vs. the nation state: Agro-food systems and the new politics of capital." *Capital & Class* 43(2): 83–106.

McNeeley, R. L. 1961. "Purse-seine revolution in tuna fishing." *Pacific Fisherman* 12(June): 27–58.

Maggs, John. 1992. "EC will protest U.S. tuna embargo against 20 nations." *Journal of Commerce and Commercial* 391(27684): 5A.

——. 1991. "Bush team feels heat over GATT tuna ruling." *Journal of Commerce and Commercial* 389(27573): 3A.

Magnusson, Paul, Peter Hong, and Patrick Oster. 1992. "Save the dolphins—or free trade?" *Business Week,* February 17, (3252): 130–31.

Marcuse, Herbert. 1964. *One Dimensional Man.* Boston: Beacon Press.

Marine Fisheries Review. 1982a. "The expansion of the Mexican tuna fleet, 1981–1984." 44(8): 25–29.

——. 1982b. "Mexico aims for large and modern tuna fleet." 44(4): 29.

——. 1979. "Dolphin mortality incidental to the U.S. tuna fishery has been reduced sharply." 74–75.

Marsden, Terry, and Alberto Arce. 1994. "Globalization, the state and the environment: Exploring the limits and options of state activity." *International Journal of Sociology of Agriculture and Food* 4:101–32.

Marx, Karl. 1947. *The German Ideology.* Moscow: Progress Publishers.

Marx, Karl, and Friedrich Engels. 1955. *The Communist Manifesto.* Ed. Samuel Beer. Arlington Heights, Ill.: Harlan Davidson.

Mathews, Jessica. 1991. "Dolphins, tuna, and free trade: 'No country can protect its own smidgen of air and ocean.' " *Washington Post,* October 18, 114:A21.

Mayer, M. 1988. "The changing conditions for local politics in the transition to post-Fordism." Paper presented at the International Conference on Regulation Theory, Barcelona, Spain.

——. 1987. "New types of urbanization and a new status of urban politics?" Paper prepared for the Sixth Conference of Europeanists, Washington, D.C., September 10–12.

Meier, Barry. 1990a. "A new storm erupts over saving the dolphins." *New York Times,* December 8:50.

——. 1990b. "Tuna company protests accusations about dolphins." *New York Times,* December 6, 140:D4(L).

Miliband, Ralph. 1970. "The capitalist state: Reply to Nicos Poulantzas." *New Left Review* 59:83–92.

——. 1969. *The State in Capitalist Societies.* London: Winfield and Nicholson.

Mingione, Enzo. 1991. *Fragmented Societies.* Oxford: Basil Blackwell.

Mishel, Lawrence, and Jared Bernstein. 1993. *The State of Working America.* New York: Economic Policy Institute.

Mjoset, L. 1987. "Nordic economic policies in the 1970s and 1980s." *International Organization* 41(3): 403–57.

——. 1986. *Norden dagen derpo.* Oslo: Norwegian University Press.

Mjoset, L., J. Fagerberg, A. Cappelen, and R. Skarstein. 1988. "The break-up of social democratic state capitalism in Norway." In *The Left in Western Europe,* ed. P. Camiller. London: Verso.

Moal, R. A. 1972. "Cooperation in tropical tuna fishing—Economics aspects of fish production." In *Economic Aspects of Fish Production,* ed. OECD, 239–56. Paris: Organization for Economic Cooperation and Development.

Morain, Dan. 1990. "U.S. told to ban tuna imports." *Los Angeles Times,* 109:A3.

Moreira, Manuel Belo. 1994. "The firm and the state in the globalization process." *International Journal of Sociology of Agriculture and Food* 4:76–95.

Murray, Robin. 1991. "The internationalization of capital and the nation state." *New Left Review* 67:84–109.

Nash, J., and Maria Patricia Fernandez-Kelly, eds. 1983. *Women, Men, and the International Division of Labor.* Albany: State University of New York Press.

National Marine Fisheries Service (NMFS). 1977. "Food fish market review and outlook." Washington, D.C.: National Oceanic and Atmospheric Administration NOAA FFSOA F-28.

National Research Council. 1992. *Dolphins and the Tuna Industry.* Washington, D.C.: National Academy Press.

Newman, Katherine. 1988. *Falling From Grace.* New York: Free Press.

Newsweek. 1990. "Swim with the dolphins: Tuna fishing will change." 115(17): 76.

New York Times. 1992a. "Pro-dolphin accord made." June 16, 141:D9.

——. 1992b. "U.S. enforces tuna embargo." February 3, 141:D3.

——. 1991. "Bumble Bee president becomes chief too." March 6, 140:D4.

——. 1990. "Judge orders tuna import ban over dolphin kill." August 30, 139:A21.

——. 1989a. "Judge extends order that U.S. protect dolphin." January 19, 138:A17.

——. 1989b. "U.S. defends law on monitoring dolphin killings." August 23, 138:A14.

Noel, A. 1988. "Action collective, partis politiques et relations industrielles: Une logique politique pour l'approche de la regulation." Paper presented at the International Conference on Regulation Theory, Barcelona, Spain.

O'Connor, James. 1989. "An introduction to a theory of crisis theories." In *Capitalist Development and Crisis Theory: Accumulation, Regulation, and Spatial Restructuring,* ed. M. Gottdiener and Nicos Komninos, 21–29. New York: St. Martin's Press.

——. 1986. *Accumulation Crisis.* New York: Basil Blackwell.

——. 1974. *The Fiscal Crisis of the State.* New York: St. Martin's Press.

——. 1973. *The Corporation and the State: Essay in the Theory of Capitalism and Imperialism.* New York: Harper and Row.

Offe, Claus. 1985. *Disorganized Capitalism.* Cambridge, Mass.: MIT Press.

Offe, Claus, and Volker Ronge. 1979. "Theses on the theory of the state." In *Critical Sociology,* ed. J. W. Frieberg, 345–56. New York: Irvington Press.

Orloff, Ann, and Theda Skocpol. 1984. "Why not equal protection? Explaining the politics of public social spending in Britain, 1900–1911, and the United States, 1880–1920." *American Sociological Review* 49(6): 726–50.

Overbeek, Henk. 1988. "Global capitalism and Britain's decline." Ph.D. diss., University of Amsterdam.

Parks, Wesley W., Patricia J. Donley, and Samuel F. Herrick, Jr. 1990. "U.S. tuna in canning, 1987." *Marine Fisheries Review* 52(1): 14–22.

Parrish, Michael. 1992a. "Pact may stop dolphin deaths in tuna fishing." *Los Angeles Times,* June 17, 111:A1.

——. 1992b. "U.S. approves pact to protect Pacific's dolphins." *Los Angeles Times,* October 9, 111:D2.

——. 1991. "Europe nearer to dolphin-safe tuna." *Los Angeles Times,* November 23, 110:D1.

——. 1990a. "Fight for 'dolphin-safe' tuna flares up again." *Los Angeles Times,* November 29, 109:D2.

——. 1990b. "Film turns tide for dolphins at StarKist tuna." *Los Angeles Times,* April 14, 109:D1.

Parsons, Talcott. 1971. *The System of Modern Societies.* Engelwood Cliffs, N.J.: Prentice-Hall.

——. 1964. "Evolutionary universals in society." *American Sociological Review* (29)3: 339–57.

——. 1960. *Structure and Process in Modern Societies.* Glencoe, Ill.: Free Press.

Perrin, William F. 1970. "The problem of porpoise mortality in the U.S. tropical tuna fishery." In *Proceedings of the Sixth Annual Conference on Biological Sonar and Diving Mammals,* 45–48. Menlo Park, Calif.: Stanford Research Institute.

——. 1969. "Using porpoise to catch tuna." *World Fishing* 18(6): 42–45.

——. 1968. "The porpoise and the tuna." *Sea Frontiers* 14(3): 166–74.

Picciotto, Sol. 1991. "The internationalization of the state." *Capital & Class* 43:43–63.

Pijl, Kees van der. 1989. "Globalization of class and state." Paper prepared for the Political Studies Association Conference, University of Warwick, England, April.

——. 1988. "Class struggle in the state system and the transition to socialism." *After the Crisis* 4: 8–21.

——. 1984. *The Making of an Atlantic Ruling Class.* London: New Left Books.

Piore, Michael J., and Charles F. Sabel. 1984. *The Second Industrial Divide: Possibilities for Prosperity.* New York: Basic Books.

Pitelis, Christos. 1991. "Beyond the nation-state? The transnational firm and the nation-state." *Capital & Class* (43): 131–52.

Poulantzas, Nicos. 1978. *State, Power, and Socialism.* London: New Left Books.

——. 1975. *Classes in Contemporary Capitalism.* London: New Left Books.

——. 1974. "The internationalization of capitalist relations and the nation-state." *Economy & Society* 3:25–46.

Prechel, Harland. 1990. "Steel and the state: Industry politics and business policy formation." *American Sociological Review* 55(5): 648–68.

Public Citizen. 1993. "Why voters are concerned: Environmental and consumer problems in GATT and NAFTA." January: 25–31.

Pugliese, Enrico. 1991. "Agriculture and the new division of labor." In *Towards a New Political Economy of Agriculture*, ed. Friedland et al., 137–50. Boulder, Colo.: Westview Press.

Quadagno, Jill. 1990. "Race, class, and gender in the U.S. welfare state." *American Sociological Review* 55(1): 11–28.

Radice, Hugo. 1984. "The national economy: A myth?" *Capital & Class* 22:111–40.

Reich, Robert. 1991. *The Work of Nations*. New York: Alfred A. Knopf.

——. 1983. *The Next American Frontier*. Baltimore: Times Books.

Rieff, David. 1993. "Multiculturalism's silent partner." *Harpers* 287 (August): 62–72.

Rockland, Steven. 1978. "The San Diego tuna industry and its employment impact on the local economy." *Marine Fisheries Review*, Paper 1313, July:5–11.

Rosier, B., and P. Dockes. 1983. *Rhythmes economiques: Crises et changement social, une perspective historique*. Paris: La Decouverte-Maspero.

Ross, Robert J. S., and Kent C. Trachte. 1990. *Global Capitalism: The New Leviathan*. Albany: State University of New York Press.

Rostow, Walter W. 1960. *The Stages of Economic Growth*. Cambridge: Cambridge University Press.

Roth, Robert. 1987. "Fordism and new social movements." Paper prepared for the Sixth Conference of Europeanists, Washington, D.C., September.

di Ruzza, R. 1988. "Les interpretations de la crise dans les theories de la regulation." Paper presented at the International Conference on Regulation Theory, Barcelona, Spain.

Sabel, Charles. 1982. *Work and Politics: The Division of Labor in Industry*. Cambridge: Cambridge University Press.

Salia, Saul B., and Virgil J. Norton. 1974. *Tuna: Status, Trends, and Alternative Management Arrangements*. Paper no. 6, The Program of International Studies of Fishery Arrangements. Washington, D.C.: Resources for the Future.

Salmans, Sandra. 1990. "Labels go green." *Marketing and Media Decisions* 15 (January): 84.

Sanderson, Steven. 1985a. "A critical approach of the Americas in the new international division of labor." In *The Americas in the New International Division of Labor*, ed. S. Sanderson. New York: Holmes and Meier.

——. 1985b. "The emergence of the 'world steer': Industrialization and foreign domination in Latin American cattle production." In *Food, the State, and International Political Economy*, ed. F. L. Tullis and W. L. Hollist, 123–48. Lincoln: University of Nebraska Press.

Sassen, Saskia. 1993. *The Global City:* Princeton: Princeton University Press.

——. 1990. *The Mobility of Labor and Capital*. New York: Cambridge University Press.

Schaeffer, Robert. 1994. "State and devolution: Economic crises and the devolution of U.S. superstate power." *International Journal of Sociology of Agriculture and Food* 4:145–58.

Schoonmaker, Sara. 1993. "State autonomy and structural constraints: Software wars in Brazil." *Critical Sociology* 19(1): 3–18.

Schor, Juliet B. 1992. *The Overworked American: The Unexpected Decline of Leisure.* New York: Basic Books.

Scott, Alan. 1990. *Ideology and the New Social Movements.* London: Unwin Hyman.

Scott, David Clark. 1991a. "Mexico chafes as U.S. revisits ban on tuna imports involving dolphin kills." *Christian Science Monitor,* February 27, 83(64): 6.

——. 1991b. "Mexico wins battle over U.S. tuna ban, but backs off to save image, trade talks." *Christian Science Monitor,* September 27, 83(213): 8.

Seafood Business Report. 1985. March/April:10.

Seafood International. 1988. "Industry in Indonesia to grow." October:35.

Selling Areas Marketing Inc. 1986. "Report for September, 1986."

Senzek, Alva, and John Maggs. 1991. "U.S., Mexico defuse tuna trade dispute." *Journal of Commerce and Commercial* 389(27562): 1A.

Sharecoff, Philip. 1990. "Big tuna canners act to slow down dolphin killings; 70% of market affected; 3 concerns will stop buying fish caught in nets that are trapping mammals." *New York Times,* April 13, 139:A1.

Skladany, Mike, and Craig Harris. 1995. "On global pond: International development and commodity chains in the shrimp industry." In *Food and Agrarian Orders in the World-Economy,* ed. Philip McMichael, 169–94. Westport, Conn.: Greenwood Press.

Skocpol, Theda. 1985. "Bringing the state back in: Strategies of analysis in current research." In *Bringing the State Back In,* ed. Peter Evans, Dietrich Rueschemeyer, and Theda Skocpol, 23–46. Cambridge: Cambridge University Press.

——. 1979. *States and Social Revolutions.* Cambridge: Cambridge University Press.

Skocpol, Theda, and Edwin Amenta. 1986. "States and social policies." *Annual Review of Sociology* 12:131–57.

Skocpol, Theda, and Kenneth Finegold. 1982. "State capacity and economic intervention in the early New Deal." *Political Science Quarterly* 97:255–78.

Stier, Ken. 1992. "Thailand cheers end to U.S. ban, prepares to release tuna stockpile." *Journal of Commerce and Commercial* 391(27684): 5A.

Strobel, Frederick R. 1993. *Upward Dreams, Downward Mobility.* Lanham, Md.: Rowman and Littlefield.

Taylor, Andrew. 1990. "Canners' 'dolphin-safe' vows spur tuna labeling bills." *Congressional Quarterly* 48(20): 1553.

Tennesen, Michael. 1989. "No Chicken of the Sea." *National Wildlife* 27(3): 10–13.

Therborn, Goran. 1984. *Why Some People Are More Unemployed Than Others.* London: Macmillan.

Thurston, Charles. 1990. "Save-the-dolphin drive to spur Asia tuna imports: Some suppliers out of stock." *Journal of Commerce and Commercial* 384(27209): 1A, 10A.

Time. 1990. "Tuna without guilt." April 23:63.

Toffler, Alvin, and H. Toffler. 1985. "An appointment with the future." *New York Sunday Times,* February 7:16.

Trachtman, Joel P. 1992. "International trade—quantitative restrictions—national treatment—environmental protection—application of GATT to U.S. restrictions on import of tuna from Mexico and other countries." *American Journal of International Law* January(1): 142–51.

Uhlig, Mark A. 1991. "U.S.-Mexico pact faces hurdle on tuna fishing." *New York Times,* April 4, 140:D14.

U.S. International Trade Commission (USITC). 1992. "Tuna: Current issues affecting the U.S. industry." USITC Publication 2547, August.

——. 1990. "Tuna: Competitive conditions affecting the U.S. and European tuna industries in domestic and foreign Markets." USITC Publication 2339, December.

——. 1986. "Competitive conditions in the U.S. tuna industry." USITC Publication 1912, October.

Vickers, Robert J. 1989. "Dolphin kill label sought for tuna cans." *Los Angeles Times,* October 5, 111:D2.

Vogel, D. 1983. "The power of business in America: A reappraisal." *British Journal of Political Science* 13:19–43.

Wallace, Charles. 1991. "Southeast Asia nations scrambling to gobble up U.S. firms." *Los Angeles Times,* August 4, 111:H3.

Wastler, Allen R. 1992. "Tuna importers struggle to escape embargo's snag." *Journal of Commerce and Commercial* 391(27661): 1A.

Weber, Max. 1983. *Max Weber on capitalism, bureaucracy, and religion: A selection of texts.* Edited and translated by Stanislav Andreski. London: Allen and Unwin.

Woods, William P., Jr. 1992. Cited in USITC (1992):63 as "vice president, Star-Kist Seafood, Co., meeting with USITC staff." Washington, D.C., March 17.

INDEX